**Tomorrow's World**

# Conference Organizing Committee

**I Arthurton**
Consultant

**D Barney**
The Engineering Link

**L Bartholomew**
Serco Railtest Limited

**D Gillan**
Railway Industry Association

**R Goodall**
Loughborough University

**R Goulette**
Adtranz

**K Herriman**
Adtranz

**R Jones**
Virgin Trains

**S Maltby**
Railway Division Young Member

**C May**
Rail Technology Services

**C Perry** (Chairman)
AEA Technology Rail

**J Smith**
Anglia Railways

**T Taig**
AEA Technology plc

**J Wragg**
Interfleet Technology Limited

# International Liaison

**Tony Roche** FEng

# Honorary Committee

**R Aylward**
Forward Trust Rail Limited

**A Baker**
Angel Train Contracts

**I Braybrook**
English, Welsh, and Scottish Railway Limited

**C Garnett**
Great North Eastern Railway Limited

**M Lloyd**
ALSTOM transport Limited

**P Staehr**
ABB Daimler-Benz Transportation

**I Warburton**
Virgin Rail Group Limited

**P Watson OBE**
AEA Technology plc

**D Wilson**
Siemens Transportation Systems

# IMechE Seminar Publication

I MECH E

## Tomorrow's World

**International Railtech Congress '98**

24–26 November 1998
National Exhibition Centre,
Birmingham, UK

Organized by the Railway Division of the
Institution of Mechanical Engineers (IMechE)

In association with the:

Institution of Railway Signal Engineers
Institution of Civil Engineers
Railway Industry Association
Unione Internationale des Transport Publics
Institution of Electrical Engineers
UNIFE
Japan Society of Mechanical Engineers

**IMechE Seminar Publication 1998–22**

Professional Engineering Publishing

Published by Professional Engineering Publishing Limited for
The Institution of Mechanical Engineers, Bury St Edmunds and London, UK.

**First Published 1998**



ISSN 1357–9193
ISBN 1 86058 187 0

A CIP catalogue record for this book is available from the British Library.

Printed and bound in Great Britain by Bookcraft (Bath) Limited.

# Related Titles of Interest

| Title | Editor/Author | ISBN |
|---|---|---|
| *New Trains for New Railways* | IMechE Conference 1998–7 | 1 86058 146 3 |
| *Train Maintenance Tomorrow and Beyond* | IMechE Conference 1997–1 | 1 86058 095 5 |
| *Better Journey Time – Better Business* | IMechE Conference 1996–8 | 0 85298 997 0 |
| *Implementing Rail Projects* | IMechE Conference 1995–1 | 0 85298 948 2 |
| *Fault Free Trains – A Reality* | IMechE Seminar 1998–8 | 1 86058 133 1 |
| *Railway Traction and Braking (Railtech '96)* | IMechE Seminar 1996–19 | 1 86058 018 1 |
| *Design, Reliability, and Maintenance for Railways (Railtech '96)* | IMechE Seminar 1996–18 | 1 86058 017 3 |
| *Railway Rolling Stock (*Railtech '96) | IMechE Seminar 1996–17 | 1 86058 016 5 |
| *Railway Engineering, System and Safety (Railtech '96)* | IMechE Seminar 1996–16 | 1 86058 015 7 |

For the full range of titles published by Professional Engineering Publishing contact:

Sales Department
Professional Engineering Publishing Limited
Northgate Avenue
Bury St Edmunds
Suffolk
IP32 6BW
UK

Tel: 01284 724384
Fax: 01284 718692

# Contents


| | | |
|---|---|---|
| C552/013/98 | **Environmental noise issues for operators of a new railway service**<br>A E J Hardy | 1 |
| C552/003/98 | **New railways – the environmental noise dividend**<br>C J Manning | 13 |
| C552/086/98 | **Design analysis of interior and exterior noise**<br>C F McCulloch and E De Geest | 21 |
| C552/063/98 | **Buying trains is easy?**<br>I R Mylroi, M J Price, A J Chaplin, and A J Walker | 33 |
| C552/070/98 | **Making light rail affordable and attractive**<br>D T Catling and T Griffin | 43 |
| C552/088/98 | **Review of the environmental conditions inside an immobilized train in a tunnel**<br>C W Pope, D J Pope, and H Barrow | 53 |
| C552/038/98 | **Steer-by-wire – a railway technology for the future?**<br>R M Goodall | 67 |
| C552/019/98 | **A new MEng Degree in railway engineering and systems**<br>D M Lilley | 77 |
| C552/004/98 | **UTILITY– utilization of transport infrastructure and land through information technology**<br>F C Perrott and A C Renfrew | 87 |
| C552/033/98 | **The use of simulators and CBT for railway driver training**<br>J W Scott | 95 |
| C552/032/98 | **Virtual reality – toy or tool?**<br>M Hayhoe | 105 |
| C552/078/98 | **Progress on the Japanese Maglev system**<br>C J Goodman | 115 |
| **Authors' Index** | | 129 |

C552/013/98

# Environmental noise issues for operators of a new railway service

**A E J HARDY**
Noise and Vibration Team, AEA Technology Rail, Derby, UK

## SYNOPSIS

Although railways are often regarded as an environmentally friendly transport system there are clearly a number of issues that concern the public. One such issue is noise, and this becomes particularly important when significant changes are made to the way trains operate. This paper examines the likely environmental impact of some of the changes a train operator may consider and suggests ways of minimising conflict with both the authorities and the local community.

## 1. INTRODUCTION

To increase the size of their business, train operators are likely to look at ways of carrying more people or freight. To attract more customers new services may well be considered. These new services might involve using:

- more trains
- "new" and "better" trains
- longer trains
- faster trains
- heavier trains

either singly or in combination. Any one of these approaches could involve changes to the noise environment near to the railway. This may be particularly true if these changes are accompanied by changes to the railway infrastructure.

The public has become increasingly aware of environmental issues. There can be few people who are not aware of the controversy resulting from the proposed High Speed Rail Link to the Channel Tunnel. It is now increasingly common to find even relatively minor changes resulting in complaints and controversy. By considering the development of a number of simple hypothetical new railway services this paper considers the public relations and legal implications in the field of noise.

In interpreting the significance of any changes, the guidelines in Table 1 are useful. However, it should be noted that changes in the character of the sound may produce a subjective response far greater than the variation in the A-weighted sound pressure level ($L_A$) would suggest. For example, if the changes involve increases in the amount of low frequency sound or introduce a tonal component to the noise levels the variation in the measured $L_A$ may be insignificant, but subjectively it could be highly significant.

**Table 1 - Approximate Significance of Changes in Railway Noise Levels**

| Change in Noise Level (SEL, $L_{Aeq}$) | Levels of Significance |
|---|---|
| Less than 3 dB | Insignificant |
| More than 3 dB but less than 6 dB | Slight |
| More than 6 dB but less than 10 dB | Moderate |
| More than 10 dB | Substantial |

## 2. PROPOSED NEW SERVICE

To attract more customers the train operator needs to offer something "new". There are a wide range of alternatives, from simply fitting new features into the existing services, to building a new railway with new trains. In between these are a wide range of possible options.
Even without building a completely new railway it may be necessary to make significant changes to the infrastructure. For example, increasing the route capacity may involve reinstating lines that have been removed, while reducing journey time may require the track to be realigned at some sites so that speeds can be raised.

When the noise impact of a new service is considered it is frequently applied only to moving trains. In practice, changes in traction equipment can be far more significant around stations and sidings. For example, simply using a new diesel engine with different noise levels may well have a significant effect on the noise environment near stations, sidings and signals.

Furthermore, the noise from an idling diesel engine tends to come from its carcass. This means that it is very difficult to control the noise levels after the trains have been built. A change in traction noise levels may have a minimal effect when the train is moving at speed, but can be highly significant when the train is standing in a station, in a siding or at a signal.

## 2.1 More Passengers or Freight on Existing Trains

At first it may appear that this offers a solution with no environmental impact. This may be true of passenger services where the weight of the train does not change to any great extent, but for a freight service the change could be significant. On freight trains the payload can make up a major part of the total weight of the train. If carrying more on the same train involves increasing the load then the locomotive will need to work harder to operate the service, with a possible increase in noise levels at some locations. A less obvious problem occurs on locomotives fitted with dynamic braking systems (rheostatic or regenerative). With heavier trains the demands on the brakes are higher, resulting in the cooling fans on the rheostatic brakes running for longer. Although the increase in traction noise is likely to be most significant on diesel locomotives, the noise from the rheostatic brakes is likely to be significant for both diesel and electric locomotives.

In addition there is evidence that heavier trains increase the re-radiated noise from under-bridges. Although this effect is small (around 1 dB to 2 dB for a doubling of the weight of the train) it may become important when combined with other changes.

Figure 1 shows how an increased load is expected to change the noise levels from a diesel-hauled train as it accelerates from rest. To obtain a proper comparison of the sound energy present at each site the Sound Exposure Level (SEL) has been used rather than the maximum A-weighted Sound Pressure Level ($L_{A,max}$). This is because the speed of the train is not constant and the time taken to pass a particular site will vary. In fact, the maximum noise would initially be independent of both distance from the start and the load. To obtain the A-weighted equivalent continuous sound pressure level ($L_{A,eq}$) over a particular period it is necessary to add to the SEL $10\log_{10}$(number of trains) and subtract $10\log_{10}$(length of the period in seconds).

Initially, just before the trains starts to move the noise levels will be the same. However, it can be seen from Figure 1 than soon after the start the heavier train tends to be noisier. This is because the noise from the locomotive dominates at low speeds and the heavier train takes longer to pass. At around 1250 metres from the start, rolling noise starts to dominate and the heavier train begins to produce a lower SEL than the train with the original load. This occurs because rolling noise is strongly speed-dependent. From Figure 1 it is clear that simply increasing the load has an insignificant impact on overall levels when expressed as SEL.

Although the data used to make the predictions for Figure 1 are typical, and the A - weighted SEL is normally used for such predictions, the impact of the locomotive noise may subjectively be greater than Figure 1 indicates. This is because diesel locomotives tend to produce relatively high levels of low frequency noise. Locomotives with effective low frequency silencers probably produce a subjective "loudness" against distance curve very similar to that shown in Figure 1. However, locomotives with poor silencers may result in levels of subjective "loudness" that result in the locomotive noise dominating for much greater distances from the start. The higher levels of low frequency sound produced by some locomotives may result in fixtures and fittings inside buildings rattling and, as a result, give rise to worries about ground-borne vibration.

Figure 1 indirectly indicates the consequences of operating locomotives in multiple formations. Clearly the train load has little influence on the total noise produced by the train.

However, Figure 1 applies only to a train with a single locomotive. Using two locomotives would accelerate the train faster, but the additional locomotive would produce an increase in the SEL of around 3 dB to a distance of approximately 650 metres from the start. This becomes even more significant when the fact that the noise produced by a locomotive is effectively independent of the size of the engine is taken into account. As a result there is clearly an environmental advantage in hauling heavy trains with a single high powered locomotive rather than with two or more lower powered units.

## 2.2 More Trains

Another relatively simple solution is to run more trains by either sub-dividing the existing trains (for example by making an 8 car multiple unit into two 4 car units), or by introducing "new" trains. Simply sub-dividing existing trains means that the total number of vehicles remains the same and consequently there will be no change in the overall noise levels measured in terms of the day or night time $L_{A,eq}$. There are a variety of ways that "new" trains could be introduced, but for the purposes of this paper the following three cases are considered:

- The "new" trains are acoustically identical to the existing trains.
- The "new" trains are quieter than the existing trains by around 8 dB(A) and only operate additional services.
- The "new" trains are quieter than the existing trains by around 8 dB(A) and operate all the services.

The difference of around 8 dB(A) is typical of that which occurs when cast-iron tread braked wheels are replaced with disc-braked wheels.

The effects of introducing "new" trains for the three cases under consideration have been calculated, assuming that the "new" and "old" trains consist of vehicles of similar length and with the same general wheel configuration. The results are shown in Figure 2. It can be seen that for the first two cases there is an increase in noise levels as the number of trains increases. The benefit of operating quiet trains only on additional services is relatively small. If, however, the service is operated entirely by quiet "new" trains, then increasing the number of trains produces a line parallel to, but 8 dB lower than, the line representing the situation where all trains are acoustically identical to those on the original service. To achieve the same noise level as the original service by operating only "new" quiet trains would require a 500% increase in the number of trains.

Clearly there is a wide range of possible combinations of "old" and "new" quiet trains that would produce the same levels as the original service. For example, doubling the number of trains and making them 40% "old" and 60% "new" results in the same noise levels as the original service.

## 2.3 Longer Trains

The day and night time $L_{A,eq}$ for identical vehicles depends on the number of vehicles. Therefore, making trains longer has the same effect on the overall noise levels as introducing more trains. By simply changing the x-axis in Figure 2 to "Increase in Train Length" it is possible to see how this parameter affects overall noise levels.

## 2.4 Faster Trains

Figure 3 shows how rolling noise levels typically increase with speed. However, if the service is operated by "new" quiet trains rather than "old" trains then the curve in Figure 3 will be 8 dB (typically) lower. This means that the "new" quiet trains could travel over 240% faster before reaching the original noise levels.
In practice a 'new' train service may involve a combination of changes in speed, number of trains, lengths of trains, etc. and the noise levels will be determined by all these factors. It should be noted that the quoted differences in noise level between disc-braked and cast iron tread-braked stock occur only on good quality track. If the rail head is corrugated or has a high level of surface roughness then the difference between disc and cast iron tread-braked stock is diminished, and in extreme cases it may disappear completely.

# 3. LEGAL IMPLICATIONS

The Noise Insulation (Railways and Other Guided Transport Systems) Regulations 1996[1] provide statutory noise protection for people living near new or altered railway lines. These regulations specifically exclude changes in noise levels that result from the increased usage of existing lines. What constitutes an altered railway is not comprehensively defined but includes tracks that have been re-laid or reopened after being closed for some years, and re-alignments. The regulations make it mandatory to insulate eligible buildings (or make a grant towards the cost of insulation) in cases where the noise levels from vehicle movements on a new system or an additional existing system reach specified triggers. The regulations apply only to moving trains and specifically exclude:

- noise from stationary trains
- general noise from station activity
- noise from shunting (defined under Regulation 2(3) as marshalling, repositioning or segregating vehicles)
- perceptible vibration
- audible ground-borne noise

The regulations divide the full 24 hours into two periods.

- day-time 0600h - 2400h
- night-time 0000h - 0600h

The regulations set three trigger levels, which must all be reached for entitlement to sound insulation. The trigger levels are exceeded when:

1. The $L_{A,eq}$ measured or predicted 1 m in front of the most affected facade of a property is not less than 68 dB in the day-time or 63 dB in the night-time.
2. The highest, predictable total railway $L_{A,eq}$ occurring within 15 years of the new or altered railway being open to traffic exceeds the prevailing $L_{A,eq}$ (day-time or night-time) by at least 1 dB.
3. Noise from vehicles using the new, additional or altered railway makes an effective contribution to the relevant $L_{A,eq}$ of at least 1 dB.

Clearly these levels can be triggered by a wide range of changes. As an example, for the following conditions:

- at the facade of a building 25 metres from the nearest line
- a service comprising 8-coach trains travelling at 160 km/h on good quality CWR
- an open site with no barriers or cuttings,

a 25% increase in the number of trains, combined with at least 20% of the total trains running on the new or altered railway, would meet all three triggers provided the original service comprised at least 25 cast iron tread-braked trains or at least 140 disc-braked trains in the 18 hour day. Alternatively, if there was no existing railway then all the triggers would be met with at least 30 cast iron tread-braked trains or 175 disc-braked trains.

The above cases are relatively simplistic. Most sites involve a mixture of trains at a range of speeds. The site may include bridges, points and crossings, jointed track, etc., and consequently each site needs to treated on its merits.

The regulations specifically exclude a number of situations. This does not mean these are not covered by legislation. For example, other parts of the Land Compensation Act (of which the Noise Insulation Regulations form a part) may be used in at least some of these situations.

If people living near a railway feel that their environment has deteriorated significantly as a result of changes to the railway they will complain. If they feel strongly enough they are likely to attempt to obtain redress through the law. Should this fail they may well enter the political arena. In a democracy it is possible, with enough political pressure, for new legislation to be developed to cover the perceived problem. For example, there is growing concern about noise levels 'creeping' upwards as might occur when the amount of traffic increases on a road or railway. If this concern turns into political action then legislation may result. Such legislation is unlikely, however, to offer much to influence the perpetrators of the original problem.

When no obvious legislation is involved there may be the temptation to think that noise issues can be ignored. This approach is at best ill advised. If enough people are upset it is not only possible for bad publicity to occur but it is also possible that a damaging court case or even new restrictive legislation may result.

## 4. PREDICTION AND MEASUREMENTS

Establishing the noise levels that existed before and after the new service is introduced is an essential part of avoiding future liability. This can be done using either predictions, or measurements, or a combination of the two. Generally, it is not really possible to carry out measurements for the "after" situation while the project is still at the planning stage. As it is at this stage that the greatest influence on future noise levels can be exerted, it is clear that some form of prediction is required.

The method commonly used to predict the noise from railways, which is also a requirement when predicting noise in connection with the Noise Insulation Regulations, is found in the Department of Transport's "Calculation of Railway Noise 1995"[2] (CRN). This document has formed the basis of the calculations presented in this report so far. However, this method does have a number of shortcomings. The most obvious is that it does not include all the rolling stock types likely to be encountered. This problem will become greater as new rolling stock is introduced.

Locomotives and power cars are a particular problem when applying CRN because it is assumed that all diesel locomotives on full power, except for the Class 60, produce the same noise levels. This is clearly not the case. The problem becomes more severe with modern locomotives that use thermostatically controlled fans and whose noise emission is therefore effectively independent of engine load. Furthermore, the cooling systems on these locomotives are designed for extreme conditions, and as a result the full cooling capacity may be used very rarely. However, CRN considers only two conditions for a locomotive; coasting, when rolling noise dominates; and full power. It has been argued that "full power" should include all the fans running. This effectively penalises locomotives with thermostatically controlled cooling fans that are on average quieter than locomotives fitted with fans that run continuously. Interpreting "full power" as "maximum noise" causes particular problems for predictions involving locomotives with rheostatic brake fans. In this case the maximum noise level will occur when the locomotive is on full power, with the engine cooling and rheostatic brake fans running. Although this situation can occur (the rheostatic brake fans will continue to run for a time after the brake has been used) it is not a common occurrence. Using the maximum noise levels in CRN not only leads to inaccurate predictions but will obscure the advantages of using modern locomotives. What is required is a prediction system that is closely related to the actual performance of the locomotives.

A similar problem occurs with the noise from under-bridges. CRN covers only three basic types of bridge, and yet within those categories there can be a wide range of sub-types, each with different noise levels. For example, CRN gives a correction for a steel bridge with ballasted track of +4 dB. In practice steel bridges can produce corrections over a range greater than 10 dB. As with the locomotives this can not only lead to significant errors in the calculations but also discourage good practice that would reduce noise levels.

With an existing railway there is clearly the opportunity to measure noise levels. CRN states how this should be done. However, it is not always possible or practicable to measure at all potential sites but, by using a combination of predictions and measurements, it is usually possible to achieve a good estimate of existing noise levels. In choosing measurement sites, consideration should be given both to those with features which cause difficulties in predicting with confidence (such as bridges), and a sample of sites that are typical. It also needs to be remembered that once the changes have been made there is unlikely to be another opportunity to take measurements of the 'before' case.

To consider the 'after' case at the planning stage it is generally necessary to make predictions. If the 'new' service simply involves the use of trains that already operate on the route then the predictions can be undertaken with a high degree of confidence. However, if trains that are new to the route are involved, then the process is more complex and has a higher degree of uncertainty. In this case it is necessary to rely on CRN or to make use of measurements taken at other sites. However, these two options are only available if the 'new' train, or something acoustically identical, already exists. When it is impossible to obtain the data on a "new" train by measurements it becomes necessary to undertake predictions based on what is known about the new vehicle. This naturally leads to a higher degree of uncertainty than would be expected when using data based on measurements.

## 5. NOISE CONTROL

The most effective method of reducing noise is to control it at source. In the case of rolling noise the source is the interaction between the wheel and the rail as a result of the combined surface roughness at their interface. It is not surprising therefore to find that reducing the combined roughness is the most effective way of controlling rolling noise. The difference between the noise radiated by trains with cast iron tread brakes and those fitted with disc brakes can be attributed to the differences in the roughness of the running surfaces of the wheels. Similarly, the differences in the noise levels produced by good quality track and corrugated track can be attributed to the differences in the roughness of the running surface of the rail. Currently, the smoothest wheels are normally found on trains with disc brakes. However, there is some evidence that some composite tread brakes may achieve similar levels of roughness. The parameters affecting rail roughness are rather more complex and less well understood. However, it is known that rail grinding can return the rails to a 'smooth' condition.

Once the noise has been generated it is usually controlled by placing some form of barrier or attenuator between the source and the receiver. For rolling noise, trackside barriers are commonly used. Although these undoubtedly do work they can replace a noise problem with a visual one. They are also only a local solution and can be expensive to install and maintain. Practical barriers are generally relatively ineffective at controlling engine and fan noise from locomotives. This is because these sources are frequently near the roof of a vehicle and consequently nearer the top of the barrier than the effective source position of rolling noise, which is in the vicinity of the wheel/rail interface.

An alternative to barriers for reducing rolling noise is to place a shroud around the bogie. In practice a shroud would need to be used with a low barrier close to the rail for maximum effect and so this solution might have to be restricted to new bogie designs operating on new railways.

## 6. CONCLUSIONS

Except in very trivial cases, noise issues cannot be ignored when planning a new railway service. If a new or altered track is involved then the provisions of the Noise Insulation (Railways and Other Guided Transport Systems) Regulation 1996 must be applied. Even when the regulations do not apply it cannot be assumed that noise can be safely ignored. A strategy to identify possible complaints and how to deal with them is essential if bad publicity and legal action are to be avoided. The public rightly expects the operators of the railway to consider noise and vibration levels when undertaking any changes. They are suspicious of anything they regard as cost cutting if it adversely affects them. However, with careful planning it is possible to avoid the majority of the problems and to be well prepared for others that are unforeseen. For example, the public may perceive that electric locomotives are quieter than diesels and so want an electric railway. However, it is relatively easy to demonstrate that modern diesel locomotives are no noisier than electric locomotives in many situations, and may offer other advantages.

Should complaints occur, a knowledge of the pre-existing conditions is invaluable. It must be remembered that complaints may arise purely because a change has occurred, and in a society that is increasingly likely to resort to litigation, it is best to be prepared. By being in a position to indicate the 'before' and 'after' situation the railway operator can demonstrate a commitment to controlling environmental impact

Measurements are undoubtedly the best way to establish existing and future noise levels (if a similar situation to that proposed is currently available). This is particularly true when modern locomotives or bridges are involved. When this is not possible, predictions based on CRN should be used. However, in some situations it will be necessary to estimate likely noise levels based on the proposed vehicle design.

The most effective way to control noise is at source. For rolling noise this means smooth rails and wheels with disc brakes. There is some evidence that some composite tread brakes may also produce smooth wheels but this requires further investigation. Noise barriers are the most common railway noise control technique but they do not work in all situations and may not always be acceptable to the public.

Although the measurement and prediction of the "before" and "after" noise levels may be considered to be unimportant and an unnecessary additional cost, it needs to be remembered that noise can become such an emotive issue that, if handled incorrectly, it can bring the most prestigious of projects to a halt. By being aware of the problems and dealing with them in an appropriate manner, the status of the project can be enhanced and considerable money can be saved.

## 7. REFERENCES


1. Statutory Instrument "The Noise Insulation (Railways and Other Guided Transport Systems) Regulations 1996", Number 428.
2. The Department of Transport, Technical Memorandum, "Calculation of Railway Noise 1995", HMSO.

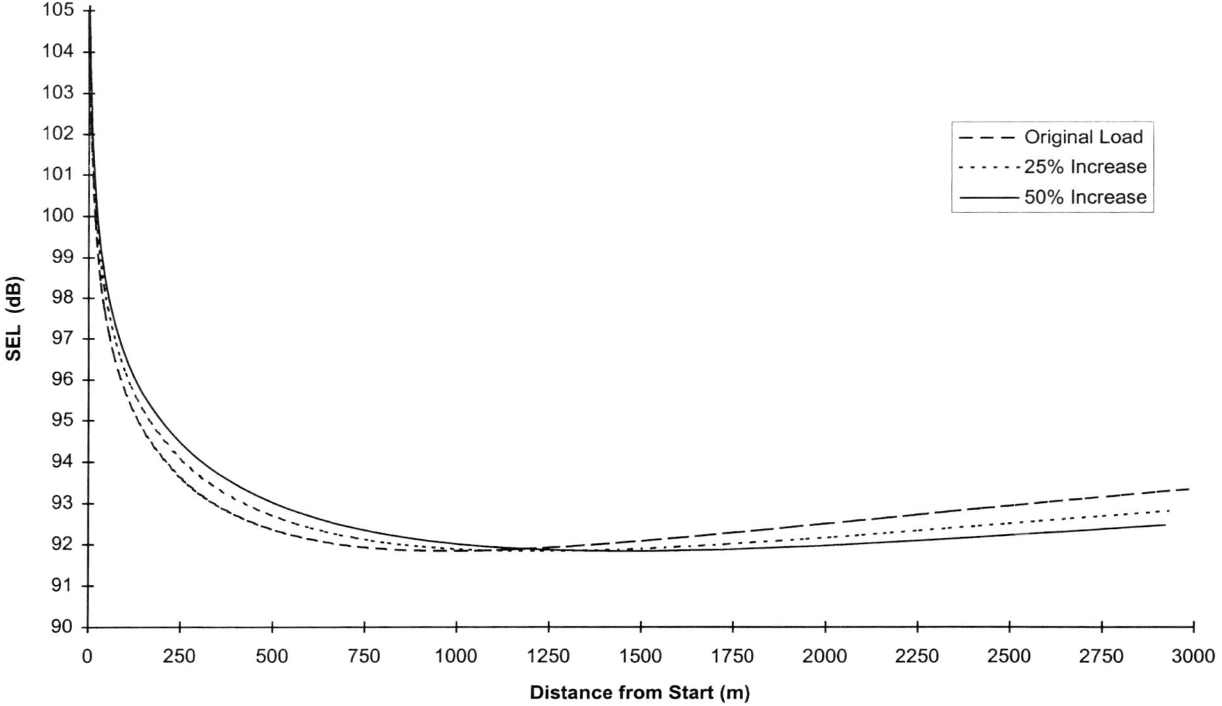


**Figure 1 - Predicted A-weighted SEL at 25 m from the Track for a Typical Freight Train with Increased Loads but the Same Locomotive, Number of Wagons and Type of Wagons**

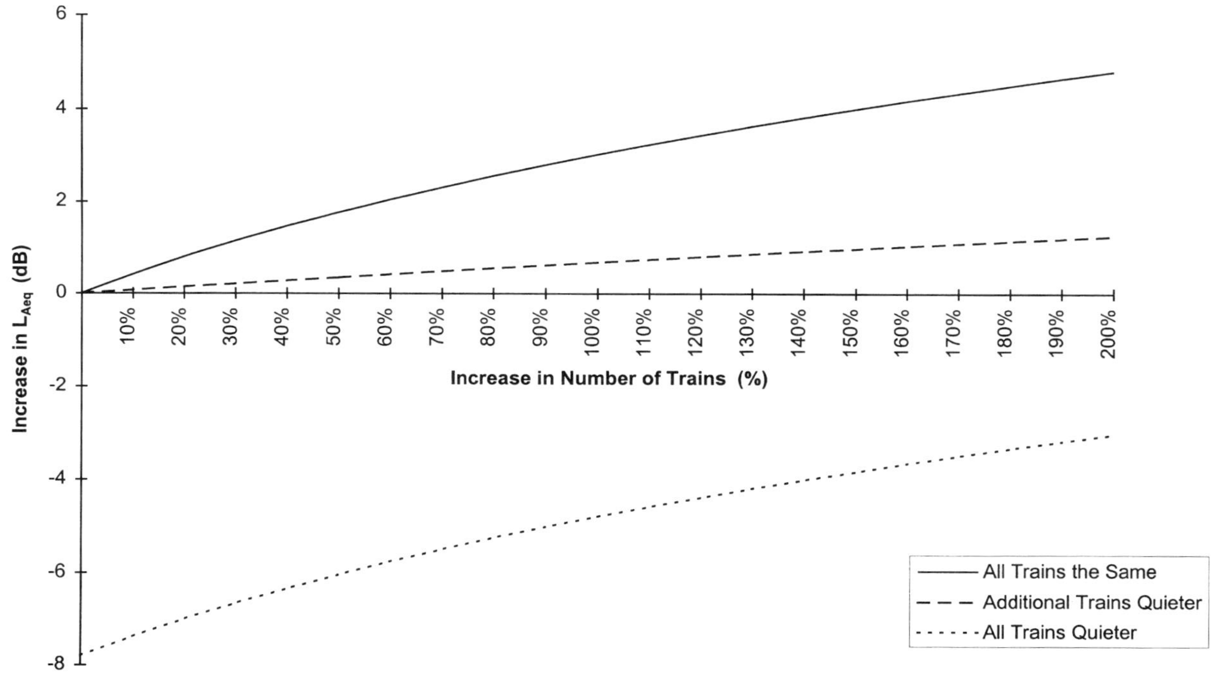


**Figure 2 - The Predicted Effect of Introducing Additional Trains on the Day-time and Night-time $L_{A,eq}$ measured at 25 metres from the track**

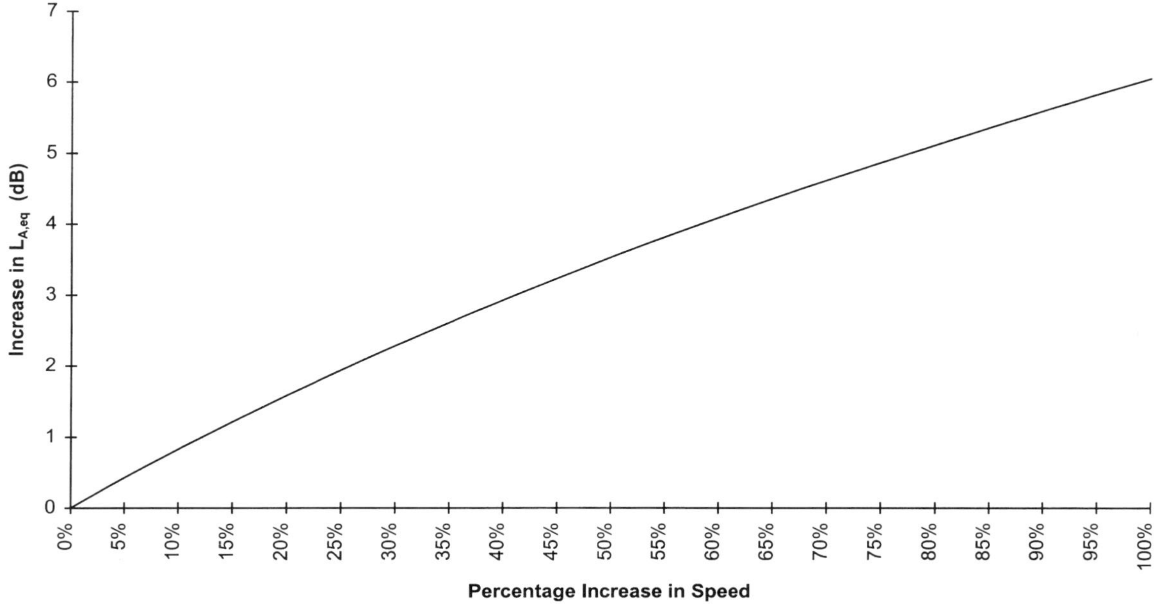


**Figure 3 - The Predicted Effect of Increasing the Speed of Trains on the Day-time and Night-time $L_{A,eq}$ measured at 25 metres from the Track**

C552/003/98

# New railways – the environmental noise dividend

**C J MANNING** BSc, MSc, FIOA, CEng, MIMechE
Arup Acoustics, Winchester, UK

## 1 INTRODUCTION

For many years in the United Kingdom we have seen the decline in public transport and, in particular, passenger and freight movement by rail. Over the same period there has been increased road building, new airports and concomitant road and air traffic growth.

All forms of mechanised transport generate some noise as a result of their propulsion systems and dynamic interaction with the medium on or through which they travel.

There are well documented increases in noise exposure from road and air transportation outstripping technical improvements in the vehicle noise generation. For road transport reduction in engine and transmission noise have been offset by increased noise from radial tyres at high speeds, acoustically absorptive road surfaces are being developed but the relentless growth in volume has outweighed all these. Similarly for aircraft low by-pass engines and low-noise flight paths have considerably reduced aircraft source noise, but again increased throughput, new runways and the growth in regional airports has increased the percentage of the population exposed to significant aircraft noise.

For railways there has been little growth or development of low-noise rolling stock until comparatively recent times. The main issues for acoustic consultants was bringing new development to existing railways, with the shortage of development land resulting in new buildings on former railway land or even in the air above railway stations.

Railway noise has only become a wider issue in the last ten years or so,seeded by a combination of legislation and general environmental awareness.

This paper describes the background to this legislation, discusses how railway noise is addressed and draws comparison with road and air transport.

Privatisation of the railway system in the United Kingdom is leading to increased investment in new railways, and rolling stock and improvements in infrastructure. The recent Government White Paper on the Future of Transport (1) proposes a new Strategic Rail Authority and increased funding for rail freight. Will the anticipated switch of passengers and freight to rail reduce the overall noise impact on the United Kingdom or will it simply transfer the impact from one sector to another?

## 2 LEGISLATION AND ASSESSMENT FRAMEWORK

Until recently, railways were excluded from normal planning and environmental legislation, being covered by an Act of Parliament. The first legislation to change this position was introduced ten years ago in response to an EEC Directive. This was the Town and Country Planning (Assessment of Environmental Effects) Regulations (2) which required, along with new roads and airport, certain new railway projects to be subject to an Environmental Assessment in the following categories:

Schedule 1: "a line for long distant railway traffic"

Schedule 2: "a tramway, elevated or underground railway, suspended line or similar line, exclusively or mainly for passenger transport".

Schedule 1 developments always require an Environmental Statement, whereas Schedule 2 developments only do if it is 'likely to have significant effects on the environment by virtue of factors such as their nature, size or location'.

A number of new railway projects have therefore been subject to Environmental Assessments over the last ten years interpreting some of the factors described above.

Assessment strategies will be covered in more detail later, but generally impacts are assessed in terms of predicted increases in noise exposure compared to baseline conditions and the significance of the effects of these impacts has regard to the severity of the impact and the number of people affected.

The other aspect of legislative control relates to compensation or mitigation offered when residential properties are exposed to noise levels in excess of trigger values. For new or altered highways, Noise Insulation Regulations were first implemented in 1973 and revised in 1975 (3) offering grants for secondary glazing and ventilation to affected residential properties. Similar schemes also exist at some airports, though these are voluntary schemes and not covered by national legislation.

In 1990 the government set up a Committee to recommend a national noise insulation standard for new railway lines which equably related to the standard set by the above regulations for new highways. This Committee reported in 1991 (4). Whilst the brief for the Committee was very specific, in order to reach its conclusions it consulted widely and based its recommendations on a full analysis of scientific evidence on the annoyance and disturbance caused by noise from roads and railways.

One of the interesting conclusions of the Mitchell Committee was that, whilst noise from railways causes more disturbance to communication than roads, in general railway noise is probably less annoying than road traffic noise with a differential of 3 - 5 dB(A) in favour of rail. This differential has been built in to some of the assessment procedures and has been referred to as the Railway Noise Dividend refered to in the title of this paper. The recommendations of the Mitchell Committee, with regard to trigger levels, were adjusted slightly, but the principal of separate day and night values was maintained. The Noise Insulation Regulations for railways were then issued in 1996 (5).

In 1994 the Department of the Environment also issued new guidance notes on planning and noise (6). Whilst intended to give guidance for new residential development next to sources of transportation and other noise it has relevance when considering the affect of new railways on land zoned for future residential development. This guidance describes Noise Exposure Categories which have the same noise level for all sources of noise at the onset but at the upper limit, above which planning permission for homes is normally refused, again shows a 2 dB tolerance in favour of rail over road and a further 2 dB over aircraft.

## 3 ASSESSMENT OF NOISE FROM RECENT NEW RAILWAYS

### 3.1 Units and terminology

It is first necessary to define and explain a few of the units and terminology used for transportation noise assessments. All transportation noise varies with time as individual sources of noise pass by a receptor point, and is also characterised by its frequency content. Fortunately there is agreement that the amplitude and frequency content of the noise can be described in a simple way using the A-weighted sound level in decibels, denoted dB(A). This weighting is applied in a measuring instrument to the actual frequency spectrum of the noise and therefore permits a single figure descriptor. The time variance is, however, dealt with in different ways for different types of noise. Railway and aircraft noise is described by the equivalent continuous sound level, denoted $L_{Aeq}$. This is a notional steady sound over an assessed time period which has the same acoustic energy as the actual fluctuating sound. For individual events one can also define the Sound Exposure Level (SEL) which, if maintained constant over a one second period, would cause the same sound energy as the actual event. Given the SEL for each event and the type and number of events one can calculate the period $L_{Aeq}$.

Other time varying sources of noise are characterised statistically. For road traffic in the United Kingdom the level exceeded for ten per cent of the assessment period, denoted $L_{A10}$, is used and environmental background noise is defined as the level exceeded for ninety per cent of the assessment period, the $L_{A90}$. Elsewhere in Europe, $L_{Aeq}$ is also used for road traffic and it is anticipated that the United Kingdom may change over to this unit soon (the guidance in reference 6 already has).

### 3.2 Environmental Statements

In order to see how the Assessment of Environmental Effects are being interpreted I have examined a number of Environmental Statements for new railway projects as follows.

**Channel Tunnel Rail Link (7):** Although published in 1994, the Environmental Noise Assessment and evaluative criteria were developed many years earlier, pre-dating the Mitchell Committee Report and Planning Policy Guidance (4 and 6). A 24-hour $L_{Aeq}$ assessment was used, though maximum noise levels were taken into consideration at night. A scale for assessing impacts was used based on predicted change between ambient noise and predicted railway noise in $L_{Aeq,24hr}$ units. The significance of the impacts was then related to the number of properties affected.

**Strathclyde CrossRail (8):** The Environmental Noise Assessment for this project was completed in 1995 and split the assessment into day and night, though in practice there were no timetabled night measurements. Reference was made to the Mitchell Committee findings and the incremental steps of noise change were all set 3 dB higher than for an equivalent road noise assessment.

**Sunderland Metro (9):** The Environmental Noise Assessment for the project was completed in 1996. Because it was partly replacing main line trains on existing track and partly all new track different assessment categories were used for impact assessment. Where ambient noise levels were controlled by road traffic, a similar approach to Strathclyde CrossRail was used, incorporating a 3 dB railway noise dividend. Where ambient noise levels were controlled by rail traffic, alongside existing lines, no dividend was applied since one was comparing like with like. There was also a third category where ambient levels were very low, and regard was had to the background noise level in setting thresholds.

**Thameslink 2000 (10):** The main noise factor of this project is a new viaduct near London Bridge Station to increase through traffic, north-south across London. Hence no railway dividend was applied but there is a significant contribution to the ambient from existing trains.

In general the assessment methods for these projects have been similar and are similar to roads where ambient conditions are controlled by rail traffic and one is assessing increments in a single dominant type of noise. Direct comparison of new railway noise with ambient road traffic controlled noise is inappropriate, unless some regard is had to the railway dividend.

**3.3 Noise insulation provision**

None of the above railways is yet operational, but some of the above assessments reported preliminary estimates of the number of residential properties that might qualify under the new regulations. These are summarised in Table 1.

There appear to be a relatively few number of expected qualifications for this cross section of project types for which the new regulations were designed. If the regulatory regime for new railways is indeed equable with that for roads, one would have expected more qualifications. Whilst this sample is small I have asked many professional colleagues who have not yet identified any residential property having been fitted with insulation under the Railway Noise Insulation Regulations.

## 4 COMPARISON OF RAIL AND OTHER MODES OF TRANSPORT

**4.1 Comparison of source noise levels**

The first comparison is between the transport options for passengers in an urban area, for which data for the Sunderland Metro Study is used as a reference. The existing options are private car, bus (diesel powered) or commuter train (diesel powered).

The corresponding SEL values based on measured data are compared to the Metro train in Figure 1. It is important to also consider the potential number of passengers carried and these are also shown in the figure. It is then possible to calculate an SEL value for each passenger carried. This may seem a strange concept, but it has parallels in looking at fuel efficiency, so allocating a noise pollution factor to each passenger also enables one to estimate how many passengers could be carried for a given noise dose at a wayside receptor. The results, in table 2, clearly show the advantages of the new Metro.

Care should be taken since the accuracy of these data are not high. Replacing diesel powered buses with electric ones would be expected to show a similar results. There is a clear indication, however, that a guided electrically powered transport system would create less noise pollution than a public or private road vehicle system using internal combustion engines.

For the second comparison I have taken a Freightliner train at 80 kmph and a single heavy goods road vehicle at the same speed and calculated the reference SEL values at 25 m distance. The results are presented in Table 3. When allowing for the relative capacity, in this case one would need some 50 road vehicles to transport the same amount as a single train, the train proves to be some 10 dB quieter than the road vehicles.

Alternatively, as with my urban passenger commuter example, one can look at the capacity of the system for a given level of noise pollution. On this basis ten times more goods could be carried by rail than road for the same noise exposure at a wayside receptor.

Again this result should be viewed with some caution because of variability between sources and, one has to consider the whole route noise exposure. For any rail freight service there are additional road journeys to and from rail hubs which add to the overall distance carried compared to a direct road door-to-door service. There is sufficient margin in the simple model, however, to conclude that there is some noise benefit from transfer of freight from road to rail.

Finally, the more complex comparison of intercity, high speed passenger services. In this case the comparison really needs to be made between rail and air services. A simple model may be very misleading because noise impacts from aircraft are concentrated around the airports during take off and landing, with virtually no impacts on-route. For the high speed passenger train the noise impact is linear along the alignment.

A simple comparison of Intercity train noise at 108 kmph, 12 carriage EMU, and 250 passenger cars at the same speed at 25 m distance to give an approximately equal passenger carrying capacity show again some 10 dB in favour of rail or ten times carrying capacity for a given noise exposure.

## 5 NOISE MITIGATION

Road and air traffic has been under pressure by legislation for many years to reduce its source noise levels. As mentioned earlier, this has been successful but offset by growth. Railways have not had this pressure and are therefore at an earlier environmental noise design stage. The sources of noise on operating railways are beginning to become well understood and new rolling stock is inherently quieter primarily due to improvements in suspension and braking systems. There is a need for more attention to the noise from locomotives be they diesel or electric, and the associated cooling systems. The main noise source for most medium and high speed trains remains rail/wheel interaction and research is continuing into this.

The use of wayside barriers on railways is also in its infancy but is potentially much more effective than on roads. Because the location of the noise source is well defined, low level absorptive barriers can be installed close to the track and many such schemes are being developed. My own company is involved in two such designs on elevated structures on the Channel Tunnel Rail Link and WestRail project in Hong Kong.

## 6 CONCLUSIONS

It is very difficult to draw accurate detailed comparisons between the environmental noise generated by different modes of transport. In this paper several different ways of looking at comparative noise levels, primarily between roads and railways, have, however shown a consistent pattern. Passenger for passenger, load for load, short or long haul there is an environmental gain to be had by switching from road to rail. Provided noise mitigation in railway track and rolling stock is developed in parallel, the increase noise from increased volumetric use of railways in the future can be offset so that the short term gain is maintained.

**Table 1: Qualification under the Noise Insulation (Railways and Other Guided Transport Systems) Regulations**

| Project | Estimated number of qualifying residential properties |
|---|---|
| Channel Tunnel Rail Link | 25 |
| Strathclyde CrossRail | 1 |
| Sunderland Metro | 0 |
| Thameslink 2000 | not known, but some anticipated due to increment on high existing train noise levels in urban area |

**Table 2: Number of passengers carried for a given noise exposure (based on Sunderland Metro study)**

| Mode of transport | No. of passengers |
|---|---|
| Private car | 1 |
| Diesel bus | 6 |
| Diesel train | 4 |
| Metro | 100 |

**Table 3: Noise Dose per passenger carried by mode of transport (based on Sunderland Metro study)**

| Mode of transport | SEL per passenger carried |
|---|---|
| Private car | 78 dB(A) |
| Diesel bus | 70 dB(A) |
| Diesel train | 72 dB(A) |
| Metro | 58 dB(A) |

**Table 4: Comparison of wayside noise for road and rail freight transport at 80 kmph**

| Source | SEL at 25 m |
|---|---|
| Freightliner loco + 25 disc braked wagons | 93 |
| Heavy goods road vehicle | 86 |
| 50 off heavy goods vehicles | 103 |

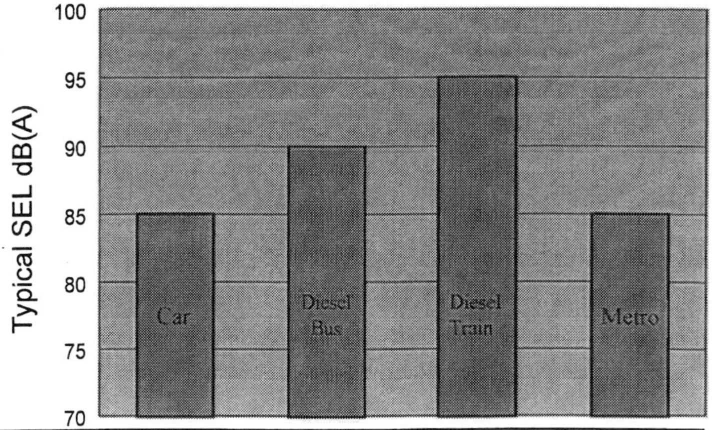


| | | | | |
|---|---|---|---|---|
| Seated | 5 | 85 | 150 | 170 |
| Maximum | 5 | 95 | 200 | 500 |

Passenger Capacity

**Fig 1. Comparison of Sound Exposure Levels at 10m for Transportation Noise Sources (from Sunderland Metro Study).**

**REFERENCES**

(1) The Future of Transport. Government White Paper, SO London, August 1998
(2) Statutory Instruments No. 1199 The Town and County Planning (Assessment of Environmental Effects) Regulations 1988
(3) Statutory Instruments No. 1763 Building and Buildings, The Noise Insulation Regulations 1975
(4) Department of Transport 'Railway Noise and the Insulation of Dwellings' (The Mitchell Committee Report ) 1991, SO London
(5) Statutory Instruments No. 428 Building and Buldings. The Noise Insulation (Railways and other Guided Transport Systems) Regulations 1996
(6) Department of Environment, Planning Policy Guidance: PPG 24 'Planning and Noise' 1994
(7) British Railways Board, Union Railways Ltd. Channel Tunnel Rail Link Final Specialist Study, Airborne Noise 1994
(8) RailTrack: Strathclyde Passenger Transport Executive: Strathclyde CrossRail Environmental Statement, March 1995
(9) Tyne and Wear Passenger Transport Authority: Sunderland Metro Environmental Statement May 1996
(10) RailTrack: Thameslink 2000 Environmental Statement November 1997

C552/086/98

# Design analysis of interior and exterior noise

**C F McCULLOCH** MA, MIOA, CEng, MIMechE and **E DE GEEST**
VPR, LMS International, Leuven, Belgium

## SYNOPSIS

This paper presents methods for modelling and assessing the acoustic environment inside and outside rail vehicles, and in railway infrastructure such as stations.
The main focus is on computer modelling using geometrical acoustics (ray-tracing) with further remarks on the use of statistical energy analysis for vehicle structures, finite element and boundary element acoustic models and some similar special methods for local models of components and construction details such as floor and trim panels.
Various examples show how the methods can be used in practice.
In the aural presentation, some synthesized (predicted) acoustic samples are replayed.

## 1 INTRODUCTION

Specifications for the design characteristics of rail vehicles, and the infrastructure (stations, etc) in which they are used, increasingly include acoustic requirements, and these requirements are increasingly expressed in terms of specific measures of performance. Not only noise levels, but also various quality parameters and measures of the intelligibility of speech and other broadcast material, are used.

These increasingly-tight specifications place increasing burdens on the designers of the vehicles and the infrastructure, especially as delivery schedules are also being shortened. The possibilities for refining the acoustic performance of a vehicle late on in its development programme, during the prototyping stage, are reduced, and the risks (indeed, cost penalties) of failure to meet specifications are increased.

As a result, it is desirable for the vehicle and infrastructure designers to have tools which enable them to assess the acoustic performance much earlier in the their work programme. This means predictive, calculation, tools are needed, rather than purely test-based approaches: testing being necessary for verification and any final trouble-shooting, and to validate the predictive methods and improve their practical use.

This paper is concerned with various such advanced predictive tools, which are being used more and more in vehicle and infrastructure design. The methods for acoustics as for other aspects of the design, can be described as moving towards the 'virtual prototype'.

## 2 ACOUSTIC TARGETS

### 2.1 General principles

Acoustic targets or specifications can be set at several degrees of complexity. The most obvious and simplest is the overall level. However, even that is not an absolute, since the non-linear performance of human hearing and the variation of its sensitivity with frequency mean that physical acoustic pressure measurements must firstly be converted to a relative, logarithmic, scale (decibels) and secondly have frequency-weighting (A, B, C, ...) applied. For noise levels with the ranges of frequency and levels commonly found in railway environments (inside and outside vehicles) the dB(A) scale is the usual overall measure, for broad-band noise (all audible frequencies summed together).

The level due to different sources, or their 'contributions' to the total, can be used to assess which sources or which paths of noise from the sources to the receivers are significant, and thus which could if possible be improved in order to have a worthwhile effect on the overall level.

The actual process of setting acoustic targets, for the insides of vehicles, is mostly driven by marketing considerations ('what environment will the customers expect or demand?'...) tempered by practical and financial constraints ('how much does this noise treatment cost?... 'how can we achieve this noise level and still have effective ventilation/required acceleration/...?'). For crew areas, employment health rules may also apply. For the external environment, targets are set by a mixture of legislative or planning requirements, similar market considerations, and safety considerations (in the case of public address, for instance).

### 2.2 Sound versus Noise

Some sound may be wanted, some not: we can refer to these two categories as 'sound' and 'noise'. Typically, the aim is to reduce unwanted noise to a minimum, but the content of that noise, such as the shape of the spectrum, is also significant: mid- and high-frequency noise can 'mask' low-frequency noise, within an acceptable overall level, whereas reducing the overall level by reducing the mid- or high-frequency content can result in an unacceptably-high proportion of low-frequency noise, giving an undesirable character (a 'rumble' rather than a balanced noise).

Wanted sound is likely to have some sort of information content: public announcements, emergency announcements, advertising or entertainment. Then, the intelligibility of that information is as much of an issue in setting a target as is the absolute level of the sound which is the information carrier. The intelligibility may be degraded as a result of the characteristics of the space into which it is broadcast, or as a result of other unwanted noise providing a masking effect.

## 2.3 Sound Quality

The human response to noise is very complex: one sound may evoke a sense of security (the 'solid' click of a door shutting) another a sense of speed or power, yet another a sense of insecurity (is some equipment malfunctioning ?) and so on. These reactions can differ from person to person, although some average or statistical measures can be attempted. It may be expected generally that noise level is related to speed (since most physical sources like wheel-rail interaction support that idea) but in some cases the reverse is the case (*eg* where noise is induced by electric drives and transmitted by structural resonances) and this can upset people who do not understand such phenomena. Other unusual noises (again, harmonics and tones from electric drives are an example, as are the unidentified bangs, rattles and whistles associated with door mechanisms, brake releases and the like) can also be disturbing to some.

Therefore the need is for metrics to characterise noise in a more complex way than just level or even broad-band spectral contents. Metrics can be devised for the significance of tonal content in a noise (does the narrow-band tone emerge more than a certain amount above the general spectrum in the same frequency range...) but a noise spectrum is frequently more complex and transient features may exist. Advanced metrics exist, under the general heading of Sound Quality, but at the present time their application seems to be mainly in the automobile and other vehicle industries (not rail). It is also not well known how these objective measures relate to subjective responses - and acquiring such information (by panel or 'jury' testing and the like) can be time-consuming and expensive. Nonetheless it is worthwhile, since the users' perceptions of a vehicle and the final delivery of a wanted product (travel from A to B in an environment that is acceptable or even attractive) can then be achieved.

## 2.4 Speech Intelligibility

The information content of speech is 'carried' by modulating the sounds of speech. Thus, if conversation between passengers is the issue, it is important to reduce the noise levels which interfere with that conversation, by masking the modulations of the speech carrier amongst other sounds of the same frequency ranges as the carrier. All these other sounds will (by definition) be incoherent with the source of the speech, unless the environment is highly-reverberant and they are echoes of the speech itself. If information transfer to the passengers (or crew...) is the issue, then often there will be several sources (loudspeakers) which will (by definition) be coherent with each other, within certain tolerances, but which will produce interference in the environment due to reflections, different path lengths from one source and another to the listener, and so on.

Speech intelligibility for passenger-to-passenger conversation can be assessed on the basis of the spectral content of the noise in the passenger environment. This is typically a simple measure such as Perceived Speech Interference Level (PSIL) which relates to the octave-band Sound Pressure Levels (SPLs) in the mid-frequencies of the audible range.

The intelligibility of public information systems or crew audio communications can be assessed by a more complex analysis and a more advanced metric, namely Speech Transmission Index (STI) which gives an objective measure of the intelligiblity of the wanted signal in the presence of its own destructive echoes and other incoherent noises. This can be used more precisely than PSIL, since it relates to a source of sound (the speaker system) which is well-defined and controllable in a way that passengers are not.

## 3 ACOUSTIC MODELLING IN THE VEHICLE DESIGN PROCESS

### 3.1 Prediction

The design process for railway vehicles frequently suffers from the lack of any real opportunity to test and refine prototypes. Therefore, it is highly-desirable to have predictive methods for interior and exterior noise that give reasonable accuracy, as early as possible in the design process. Computer-based predictive models offer this possibility. Where they have a rational basis in the physics of the vibro-acoustic behaviour, they are likely to give more reliable and more useful results - not only in well-known situations, where they can be validated, but also in other, novel, applications.

### 3.2 Diagnosis

Once rational predictive models are established, they can be used to extract information about the reasons for the predicted results. What paths contribute to the noise penetrating to the interior? Which sources are dominant and therefore most-usefully treated? Which panels of the vehicle are significant in allowing air-borne or structure-borne noise to penetrate? Numerical results form computer models enable logical ranking and cost-benefit assessments.

### 3.3 Re-design and optimisation

The process of re-designing, adjusting details, materials, and optimising the design can be driven by 'tests' on the 'virtual prototype' formed by the computer models. Possible design changes can be assessed rapidly. The impact of changes in specifications for the inputs (motor specifications, revised ancillary equipment, ventilation, ... as often occur during the design period) can be handled. Numerical optimisation procedures are available which allow the design to be steered to a target, which can be a combination of costs, weight, performance and passenger environment. This is already being used in some applications, where the optimisation is based on a localised set of parameters (*eg* thicknesses different panels and materials, versus transmitted sound, ...) and we can expect to see this increasing in use and becoming more generalised to the whole vehicle.

 

### 3.4 Trouble-shooting

Having design models available enables rapid assessment of short-comings which may finally be reported during the verification tests. The diagnosis, re-design and optimisation processes can be activated to respond speedily to the problem, in conjunction with (and steering the direction of) re-work and re-test actions, saving time and cost in remedial work.

## 4 ACOUSTIC MODELLING TOOLS

### 4.1 Geometrical acoustics

Geometrical acoustics uses a ray-tracing method to calculate the way sound is distributed. (More precisely, beam-tracing is used, which gives more robust results). The method can be used for completely-closed interior spaces, partially-open ones and fully-open exterior environments. Each source is located in space and given a defined power. It may radiate equally in all directions, preferentially in some (by a defined directivity, *eg* a directional loudspeaker) or wholly within a defined region (eg a vibrating panel, radiating on one side only). Each source may also be basically coherent or incoherent. The rays (beams) from each source are traced around the defined space and reflect off all the surfaces they contact like mirrors (see Figure 1) losing energy at each reflection in accordance with the absorption properties of the surfaces. The rays 'capture' receiver points (occupants head locations, ...) as their cross-sections pass by, defining an echo with a certain arrival time (due to its path length from the original source) and a certain energy level (due to the energy absorbed at the reflections and the spherical spreading-out of the beam cross-section).

The result is firstly an echogram (or impulse response) at each receiver point (Figure 2). From this, many sound parameters can be derived: steady-state SPL, acoustic quality measures and STI (see [1,2]). All of these can be frequency-dependent, since both absorption properties and source powers and directivities can vary with frequency. Figure 3 shows an example of a noise map in a vehicle cabin, due to certain equipment noise. By superimposing many different sources, varying their properties, and so on, with intelligent data-handling in the modelling process, many different possible scenarios can be assessed and the effects of different combinations of sources (wanted or unwanted) can be rapidly analyzed. Typically, the basic model data such as geometry are loaded more-or-less automatically from a general-purpose CAD system, but the more-specialized acoustic data such as loudspeaker properties are usually derived elsewhere and saved in dedicated databases.

Special techniques exist for the modelling of diffraction (*eg* over track-side barriers or around shields) and the transmission of sound through partitions from one region to another (*eg* from outside to inside of the vehicle, from one compartment to another, ...). These techniques enable a complete model of the noise environment around as well as inside the vehicle to be built up. More details of the method and the procedure for auralizing sound using the results can be found in [3] and [4].

## 4.2 Statistical Energy Analysis

Geometrical acoustics is more-or-less limited to purely acoustic calculations, such as sound systems, and noise distribution in or around a vehicle. Where the noise is mainly structure-borne, it is necessary to consider mainly the structural vibration behaviour, rather than the geometrical distribution of the sound. At medium to high frequencies (which will be defined more-precisely later) the standard approach for modelling has become Statistical Energy Analysis (SEA). In this approach, the structure is broken-down into sub-systems, each of which has defined boundaries, geometry (curved shell, acoustic space, ...) properties (thickness, honeycomb core, beam cross-section, ...) and connections to other sub-systems (point, line, area, ...). The geometry, properties and connections define the vibrational characteristics of the sub-systems and their interactions. An important input is also information about damping ('loss factors') within the sub-systems and in their connections. The model, consisting of the assembly of all the sub-systems, is driven by energy inputs related to the physical sources of vibration and noise, such as suspension mounts, motors, ancillary equipment, and so on, and an energy-balance (or flow) is derived. The results are then the energy levels (vibrational or acoustic) in the sub-systems, and the flow of energy from the sources into the dissipating systems.

SEA provides an effective modelling tool for the global behaviour of structural vibrations and noise radiation to the exterior and into the interior spaces, with the benefit of being reasonably fast and therefore practical at the design stage. By integrating experimental vibration measurements on prototypes, previous designs, or sub-systems of a new design, the data input to the models can be validated and improved. Whilst an SEA model is essentially just a network of sub-systems and connections, it can also be visualised in a geometrical way: Figure 4 shows a typical SEA model of a railway car.

A special form of SEA model is used in the analysis of wheel-rail interaction ('rolling') noise, typically using the 'Remington' model based on the effective roughnesses at the contact, static load, unsprung mass, wheel disc and rim dynamics, rail mass per unit length, support stiffnesses, sleeper,tie and track-bed forms. Several proprietary computer algorithms exist for this methodology, but limited space precludes further detailed discussion here. (See for instance [5] and [6]). The results of such a model (dynamic forces into the track-bed, sound power of the wheel-rail system *vs* frequency...) can be input to the ray-tracing and structural SEA models already described as additional energy sources.

## 4.3 Finite Element and Boundary Element Methods

Finite element (FE) methods are well-established for structural models, including dynamics as well as stress analysis, but in the context of railway vehicles only low frequencies can be handled (typically up to 50 or 100 Hz maximum depending on size of structure). Beyond these frequencies, the modal density of the structural dynamics becomes too high and a deterministic method such as FE is not valid. (SEA methods can then take over). Likewise, an upper limit of 100 Hz or so also applies for acoustic FE or boundary element (BE) models of vehicle interior spaces, which may be coupled to structural FE models to take into account fluid-structure interaction (vibrating panels radiate noise, but their vibration is altered by the reaction from the air as they move...). If such models are limited to local regions, such as specific panels of the vehicle body or a small compartment or section of it, they can be useful in predicting and diagnosing local problems and the vibration and noise paths which cause them, and can be used to higher frequencies. FE structural models and FE or BE acoustic models are also very useful in the design and optimisation of components and sub-systems, such as wheels, motors/engines, and ducts such as intake and exhaust systems. Figure 5 shows an acoustic BE model of a wheel used in research to minimise the noise-radiating behaviour.

### 4.4 Vibro-acoustic panel and trim modelling

Special vibro-acoustic models using FE methods can be applied to the design of panels and trim assemblies. These models use both structural and fluid elements for the appropriate parts, and special elements with properties related to both fluid (air) and structural 'phases' in the material. Typically, these elements use Biot's method [7] to define the interaction of acoustic waves and structural vibrations, in porous or fibrous material. Thus one can model the behaviour of a structural panel, where the vibrations are altered and the acoustic radiation or transmission is changed by the addition of multiple layers of damping and absorbent material (see Figure 6). This can be especially useful in applications such as the design of floor panels or doors for minimum acoustic transmission.

## 5 APPLICATIONS

### 5.1 Noise inside a railway vehicle

An engineering model for the noise inside a vehicle can be built up using the ray-tracing approach or the SEA approach or a combination. The ray-tracing model is particularly useful for dealing with the largely airborne components of noise at mid- to high-frequencies, resulting from the wheel-rail noise, aerodynamic noise and drives and ancillary components. The possibility to include transmission through panels is especially useful, because the exterior noise field can be included in a single model: noise from sources such as the wheels propagates externally, reflects of the vehicle body (including shields and the like) and track-side objects such as barriers. Diffraction algorithms can be applied to account for the non-direct path of the noise. The incident energy on the vehicle exterior then propagates to the interior in accordance with the transmission properties and areas of the panels, creating secondary sources which re-radiate to the interior. The interior noise can then be calculated with the same model, including dispersion effects which may localise the noise to certain regions, even within an open compartment. Additional interior sources can be added, to include ventilation noise and other equipment which radiates directly to the interior. The effectiveness of acoustic panel treatments, for either absorption or reduced transmission from outside, and changes to source powers (treating equipment noise at source, adding shields, ...) can then be investigated: 'what if' comparisons are straightforward once a basic model is established. Figure 3 shows the interior noise propagation using such a ray-tracing model.

The local behaviour of specific panels can be calculated with the multi-layer modelling methods and the results of these models incorporated in the ray-tracing or SEA models, where panels such as the floors over powered bogies are identified as problems, with the aim of achieving optimum performance for minimum weight within defined size constraints.

## 5.2 Noise from a railway vehicle

The noise propagating to the exterior can be investigated by the same ray-tracing model as the interior noise, or very similar: it is the exterior model used to establish incident noise levels on the vehicle envelope, but now used to establish the power radiated to the far-field - for instance, to the standard locations for the pass-by noise specifications. In this case, though, the diffraction effects of barriers must definitely be included: for incoherent sources (such as wheel-rail and aerodynamic noise) a frequency-dependent formula using path-length differences for the diffracted *vs* direct (through-the-barrier) paths can be used, with the ray-tracing program automatically finding the relevant path lengths with minimum differences. Figure 7 shows a contour plot of track-side noise from a high-speed train.

## 5.3 Noise in and from tunnels

It has already been mentioned that the noise on the exterior envelope of the vehicle can be calculated with ray-tracing, including the effects of diffraction around the vehicle body, shields, *etc* and the effect of reflections from track-side objects. This reflection effect becomes most apparent when calculating the noise inside the vehicle when in a tunnel, where there is usually a highly-reverberant environment.

The ray-tracing method can also be applied to the environmental noise propagating from the tunnel portal, including diffraction effects around the portal sides. The reverberant effects near the mouth of the tunnel can result in a higher-than-normal effective power for the passing train as it enters or leaves the tunnel, thereby concentrating the environmental impact in the vicinity of the portal. The sudden transient effect from the emerging train can also be investigated by considering a series of 'snap-shot' time instants during that short interval. Figure 8 shows an example of a contour plot of noise levels around a tunnel portal.

## 5.4 Public address in a railway station

The use of ray-tracing to assess speech intelligibility has been described, and this is useful in assessing the performance of public information systems in the railway infrastructure, such as stations, and even on-board the trains themselves. Figure 9 shows a contour plot of Speech Transmission Index in a railway concourse. A small number of directional loudspeakers was used. The placement, orientation, power and directional properties of the loudspeaker system can be optimised, to take account of the self-destructive effects due to the large reverberant space and the interference of other sources such as machinery and background noise, for a range of scenarios in terms of occupancy, emergency announcements directed to specific zones, and so on.

 

## 6 CONCLUSIONS

Many different tools exist for predictive and diagnostic modelling of noise and vibration in railway vehicles and infrastructure. The increasing speed of the computer-based modelling means that more design options and more aspects of the design can be evaluated within the allowable time. These modelling tools can be integrated with CAD and other design methods to form a process which enables optimum design, to meet increasingly-strict targets for both passengers, crew and environmental impact, within increasingly-tight constraints on cost and time.

## REFERENCES


[1] J.Martin *et al*, *J Acoust Soc Am*, 94 (6) 1993

[2] H.Kuttruff, *J Audio Eng Soc*, Vol 41, No 11, 1993

[3] T.Houtgast *et al*, 'Predicting Speech Intelligibility in Rooms from the Modulation Transfer Function, Part I' *Acustica* 16 p60, 1980

[4] R.Plomp *et al*, 'Predicting Speech Intelligibility in Rooms from the Modulation Transfer Function, Part II' *Acustica* 46 p75, 1980

[5] P.J.Remington, 'Wheel/rail rolling noise, parts I and II' *J Acoust Soc Am*, 81 p1805, 1987

[6] D.J.Thompson, 'Wheel-rail noise generation, parts I-V' *J Sound and Vib* 161 p387, 1993

[7] VIOLINS ® Revision 5.3, LMS International, Leuven, Belgium, 1997


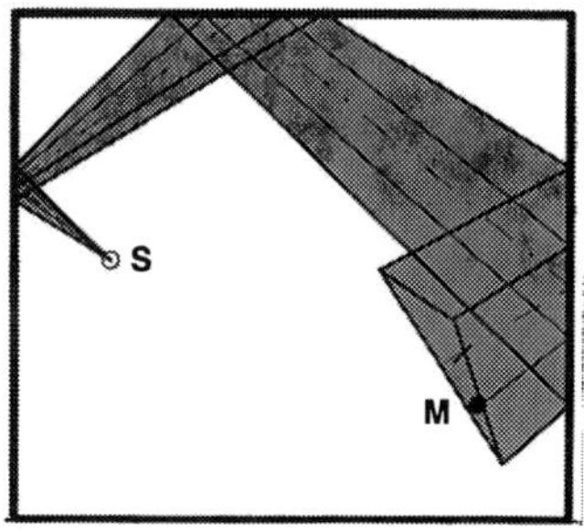


**Figure 1: Beam-tracing principle**

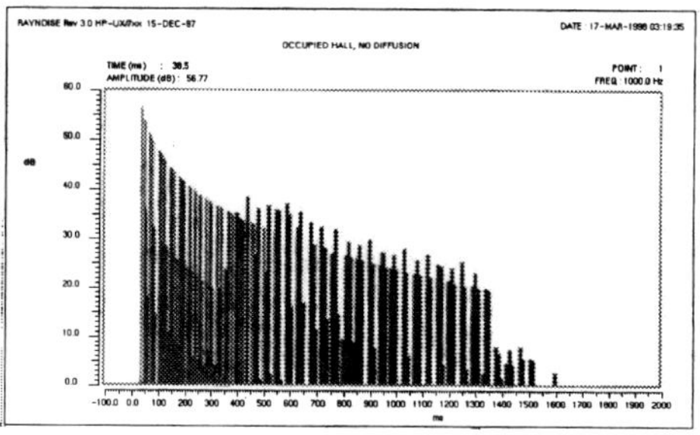


**Figure 2: Resulting echogram**

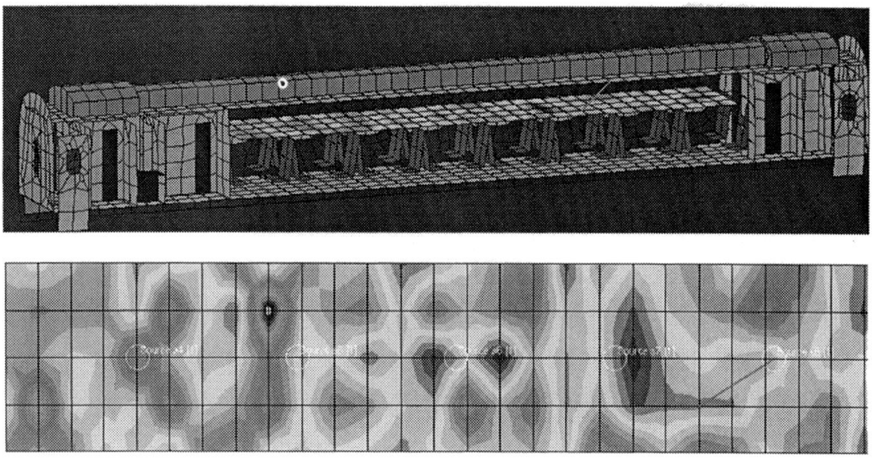

**Figure 3: Noise map inside railway vehicle**

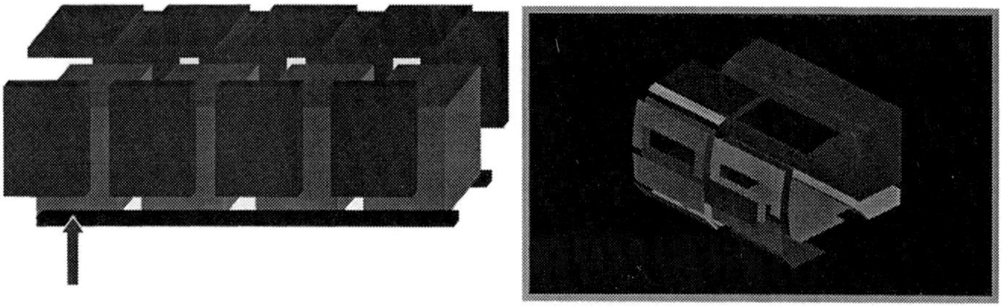

**Figure 4: Statistical energy analysis vibro-acoustic model (left, global; right, detailed)**

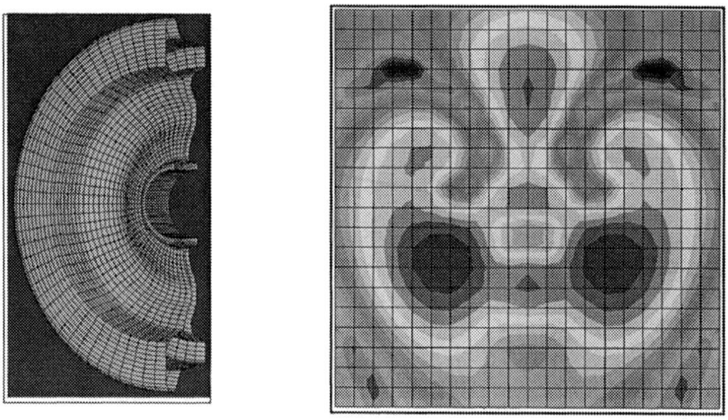

**Figure 5: Investigation of wheel noise radiation using BEM (left, mesh; right, radiated noise pattern)**

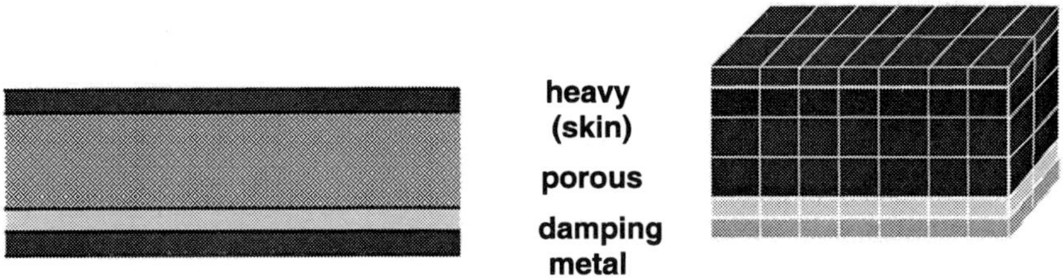


**Figure 6: Multi-layer material (floor, door, …) using poro-visco-elastic FE model**

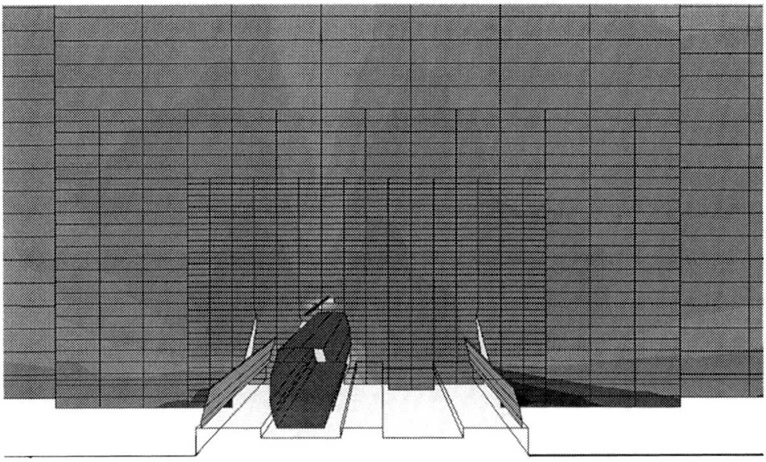

**Figure 7: Pass-by noise from train with barriers: ray-tracing model with diffraction**

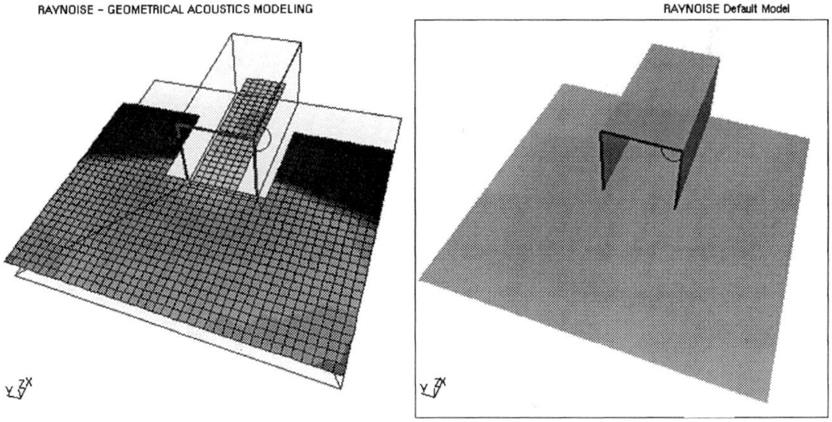


**Figure 8: Noise break-out around tunnel portal: ray-tracing model with difraction**

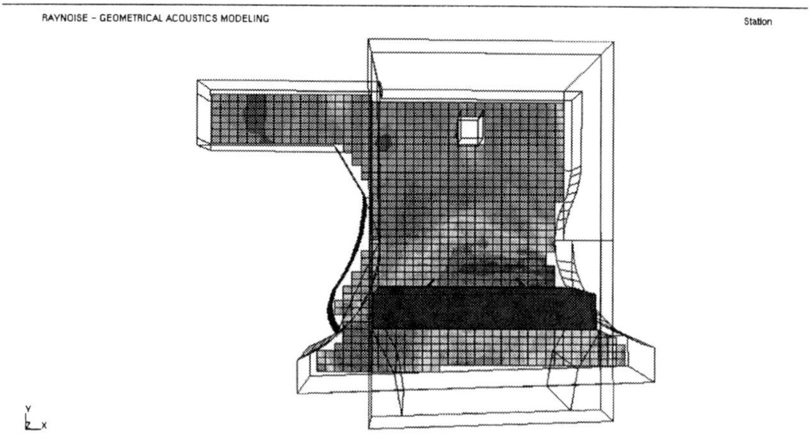


**Figure 9: Effectiveness of sound system in station concourse: phase ray-tracing model**

C552/063/98

# Buying trains is easy?

**I R MYLROI** BSc, CEng, MIMechE
Interfleet Technology Limited, Derby, UK
**M J PRICE** MBA, BSc, CEng, MIEE, and **A J CHAPLIN** MIMechIE
ScotRail Railways Limited, Glasgow, UK
**A J WALKER** BA, MCIPS
First Procurement Associates Limited, Derby, UK

## BUYING TRAINS IS EASY?

### Synopsis

Buying trains is as easy as you make it, however, committing over £100 million is never easy!

ScotRail Railways have placed orders for fleets of new multiple unit trains, some of which will be electrically powered and others diesel.

Both transactions include a bespoke maintenance support package from the train manufacturers.

The orders were negotiated and concluded simultaneously in a period of some 22 weeks; a significant improvement on the industry norm.

This paper outlines how strategic planning laid the foundation for this success.

The paper outlines how the various, competing train design, maintenance and financing options potentially available to ScotRail Railways were evaluated to deliver a solution which met the Business objectives. In so doing, the paper identifies the roles and contributions of the train manufacturers and the rolling stock leasing market.

The paper concludes that through goal-focused objectives and wider team working, significant improvements in the train procurement process are achievable.

## 1. BACKGROUND

When National Express Group (NEG) acquired the ScotRail Railways (ScotRail) franchise in March 1997, commitments were made to replace the aged train fleets operating on suburban services in Glasgow, and to increase the frequency of the high speed shuttle service between Edinburgh and Glasgow. In both cases, new, multiple unit trains were required - electrics for the Glasgow area and diesels for the shuttle service.

The franchise obligation for full fleet service introduction for both fleets is 1 April 2000; some 33 months from franchise award. During which time a competitive procurement process was to be pursued in line with European Union Procurement Legislation, suitable trains identified, lease and maintenance arrangements concluded, trains produced, Safety Cases obtained, and introduction of the trains with a low probability of a customer confidence damaging period as reliability of the units in service was increased. Such processes in total, have, in recent years taken considerably longer than 33 months, with the time to contract award alone typically taking up to 18 months of the period.

ScotRail faced a dilemma. Their franchise commitment stated that contracts for both electric and diesel multiple units (even with the granting of an extension by OPRAF) had to be in place by 31 March 1998. As ScotRail was franchised at the end of the privatisation process, obtaining new trains in such short timescales would be difficult. This problem was compounded by production capacity at the likely manufacturers being reduced as other Train Operators were ordering new trains for production in 1998/99. To stand any chance of meeting their franchise commitment to OPRAF and their customers, ScotRail had to move quickly to ensure their order(s) were placed and at the most competitive prices.

## 2. THE CHALLENGE

With their successful acquisition of the ScotRail franchise, NEG then appointed Alastair McPherson as Managing Director. McPherson brought with him considerable experience and commercial expertise to drive forward NEG's Business aspirations and meet their OPRAF obligations. McPherson recognised that if ScotRail were to have any chance of hitting the OPRAF deadline something radical needed to be done in terms of improving the process and providing dedicated resources to deliver the requirement.

The remit which emerged was to appoint consultants who knew the procurement process for new rolling stock, understood the issues and who could introduce the necessary innovation. Following a competitive process, a joint team of First Procurement Associates Ltd (FPAL) and Interfleet Technology Ltd. were appointed as professional advisers to ScotRail to assist in the procurement process.

## 3. THE NORM

The industry norm to tendering would have been against bespoke terms and conditions of contract (the Agreement), including technical specifications for both supply and maintenance of the trains.

Tenders would have been issued to pre-qualified manufacturers and once a capital purchase price was established, financiers would then have been invited to compete to become lessors.

The normal process would have dictated lengthy discussions, meetings and clarification of responses before any recommendation was made to move to a short list of manufacturers. Normal timescales for such a process was anything up to eighteen months. After this, detailed negotiations would have commenced with one or two manufacturers before a recommendation was made on the preferred manufacturer. Recent procurement exercises have seen a lessor appointed via a later, competitive process after the decision has already been made by the Train Operator on the choice of manufacturer. During and after this lengthy process, the lawyers would be invited to conclude the legal documentation, weakening the commercial position of the buyer and adding further delay to order placement.

## 4. THE STRATEGY

After meetings with ScotRail, FPAL and Interfleet developed a procurement strategy which reflected the key requirements of the client which were:

- maximise effective competition with the suppliers best able to compete
- meet EU legislative requirements
- meet the timescales
- meet OPRAF's requirements
- meet NEG/ScotRail Business objectives
- build on the experience from the Midland Mainline procurement
- process

A key lesson learned from recent procurement processes was the reason why others had been delayed. The main reason for this seemed to be that the preferred supplier was declared before the product offering was finalised and contracts were agreed. This led to negotiations becoming protracted and difficult before contract documentation was completed to the satisfaction of all parties.

If ScotRail were to meet their Business objectives, their best chance of success was by the procurement of low risk "off the shelf" products. For the potential financiers, the residual value risk would be lessened by the choice of such products which, by the time of delivery to ScotRail, would have become proven elsewhere in the UK.

ScotRail's procurement advisers, proposed to appoint the financier(s) first. As the lessor would be the subsequent asset owner, (for over 30 years), ScotRail believed that the financier should be involved and share ownership of deciding the manufacturer(s). Therefore, a joint decision between ScotRail and the financier was agreed to be the appropriate way forward.

This required the agreement of the manufacturers for ScotRail to make tender information available to preferred financiers to assess and assist with the evaluation process to jointly determine the preferred manufacturer(s).

Before and during the tender stage the procurement strategy was openly declared and shared with the supply side. Bidders were made aware from the outset of ScotRail's approach in terms of tendering, evaluation, initial appointment of a financier(s) and contract placement.

This procurement strategy was totally innovative and potentially saved many months of effort for everyone involved.

## 5. THE PROCESS

Through a detailed pre-qualification process, three DMU manufacturers, four EMU manufacturers and four potential financiers were identified.

Each was invited to provide quotations against a brief, performance specification. This detailed the aspirations of ScotRail in terms of operation and maintenance and identified key commercial and technical issues which both the manufacturers and financiers were asked to address in their responses. No bespoke terms and conditions were issued; bidders were invited to offer their standard products, together with their standard conditions of contract.

As previously stated, ScotRail were late into the privatisation process. The challenge to ScotRail and the procurement advisers was to turn this disadvantage into an advantage. ScotRail knew that other Train Operators had recently acquired new rolling stock via a financier and manufacturer though not necessarily in that order. Therefore, similar outputs had been achieved and were proven within the more commercially aware industry. Building upon this position during the tendering period, ScotRail requested both financiers and manufacturers to provide each other with information in terms of product, price, terms and conditions of contract to enable firm prices to be offered to ScotRail.

This approach is a radical change from previous procurement processes and represented the major risk to ScotRail. However, when the tender responses were received on 8th January 1998 they were sufficiently robust in content to enable ScotRail to move quickly forward with the evaluation and identification of the preferred financiers.

Bids were received from two DMU and three EMU manufacturers and from three financiers.

## 6. THE APPROACH TO THE EVALUATION

The first stage was a "rude and crude" evaluation of the bids received with particular emphasis on the financiers. This was carried out by the ScotRail project team which also included the input of senior legal and financial support from NEG. The strategy was to shortlist two potential financiers for each product required and, through a competitive process, seek to appoint one financiers who would work with the ScotRail's team in determining the preferred manufacturer(s).

The dominant criterion in this process was to ensure that the evaluation for determining the financier took into consideration not only £ per million of capital finance, but also the ability to work with ScotRail, ensuring an understanding and ability to comply with needs over time in terms of the product, strategy, timescales and contracts.

Failure to make the right choice in terms of financier would have meant that ScotRail would not have achieved its objective of the latest of 31st March 1998 for contract award.

## 7. THE OUTPUTS

Following the appointment of the financiers on the basis of Porterbrook Leasing Ltd for DMU's and Forward Trust Rail Ltd for EMUs - a series of meetings were held with the competing manufacturers to assess the products available against the Business objectives.

For the DMU, it was quickly identified that the product supplied by ADtranz best met the requirements of both ScotRail and Porterbrook in terms of product, price and time. This product was already developed from vehicles currently being manufactured by ADtranz for a sister company within NEG. Therefore, and in view of the timescales, there was obvious synergy in procuring this vehicle from ADtranz.

For the EMU process, detailed discussions were undertaken jointly by ScotRail and Forward Trust's project teams, initially with all three manufacturers, then shortlisted to two manufacturers. The process required a move to detailed and competitive negotiations around the contract terms provided by the manufacturers in their tender response. Terms included the 'boilerplate' protections and the extent to which the manufacturers would 'back' their product against ScotRail's objectives. Discussions were undertaken with the manufacturers commercial and technical specialists, together with their legal representatives. The objective of the discussions was to document the discussions as they moved forward, thus ensuring that both ScotRail and Forward Trust knew the exact position of the competition. Whilst resource hungry, it enabled pressure to be maintained on the negotiating teams from the manufacturers to ensure the most beneficial terms were secured.

Price at this stage in the process was not discussed!

Once a stage had been reached where each offer was known and clearly understood against each of the various contracts (i.e. supply and manufacture, maintenance, master lease and spare parts contracts) was the appropriate time for final detailed negotiations (including price) to commence with the manufacturers. These negotiations resulted in GEC ALSTHOM Metro Cammell Ltd offer being accepted and contracts being signed shortly afterwards.

## 8. OPRAF

In parallel to all this ScotRail kept OPRAF briefed of where they were in the process. Early on ScotRail met OPRAF and formally agreed dates when draft documentation would be provided to them for comment. The reasons for this were that OPRAF through the Direct Agreement would act as guarantor to the manufacturer/financier in the event of ScotRail defaulting, terminating the Agreement or going into liquidation. Therefore OPRAF required input into the documenting process. Also, to assist OPRAF in ensuring their legal/commercial resources were available to review and respond to documents in accordance with the agreed timescales.

These dates were met by both ScotRail and OPRAF which resulted in contracts being committed on 21st March 1998, a period of 22 weeks from inception of the strategy.

## 9. THE CONCLUSIONS

The conclusions from this process were:

- the value of spending time developing the strategy at the outset was proven;
- the proposed solution was essentially one to meet the needs of a Train Operator with a short, initial franchise period;
- the strategy was successful and, as such, provides a model for others;
- the positions and role of the supply side - particularly the pivotal role of the financier - needs to be understood and reflected, whilst maximising the benefits of effective competition;
- technological advance is potentially stunted by adoption of the ultra low risk approaches demanded by short franchise directions and commitments.

## 10. KEY MESSAGES TO ENSURE BUYING TRAINS IS EASY, OR AT LEAST MADE EASIER!

In summary the suggested points for consideration when buying new trains are:-

- **Planning** : Set realistic milestones and timescales for the process. Consistently review the plan against actual.
- **Understand**: the products, inherent risks, marketplace for manufacturers and potential lessors.
- **Procurement Strategy**: Prepare and agree strategy. Once agreed keep to it!

- **Resources**: Do not underestimate the resources required. Dedicate key individuals.
- **Teamwork**: find people with the right technical and process skills and who can be dedicated to the project.
- **Use specialist advisers**: Be clear what you want them to do and ensure that they have the skills and the culture to work with you.
- **Communication**: Consistently work with the supply side to ensure resources are available.
- **Empowerment** Ensure individuals are empowered to make decisions, but do ensure that all necessary approvals (e.g. from Boards) are obtained when they are required.
- **Reasonable and Professional Approach**: Remember you are going to be dealing with both financier and manufacturer for a long time!

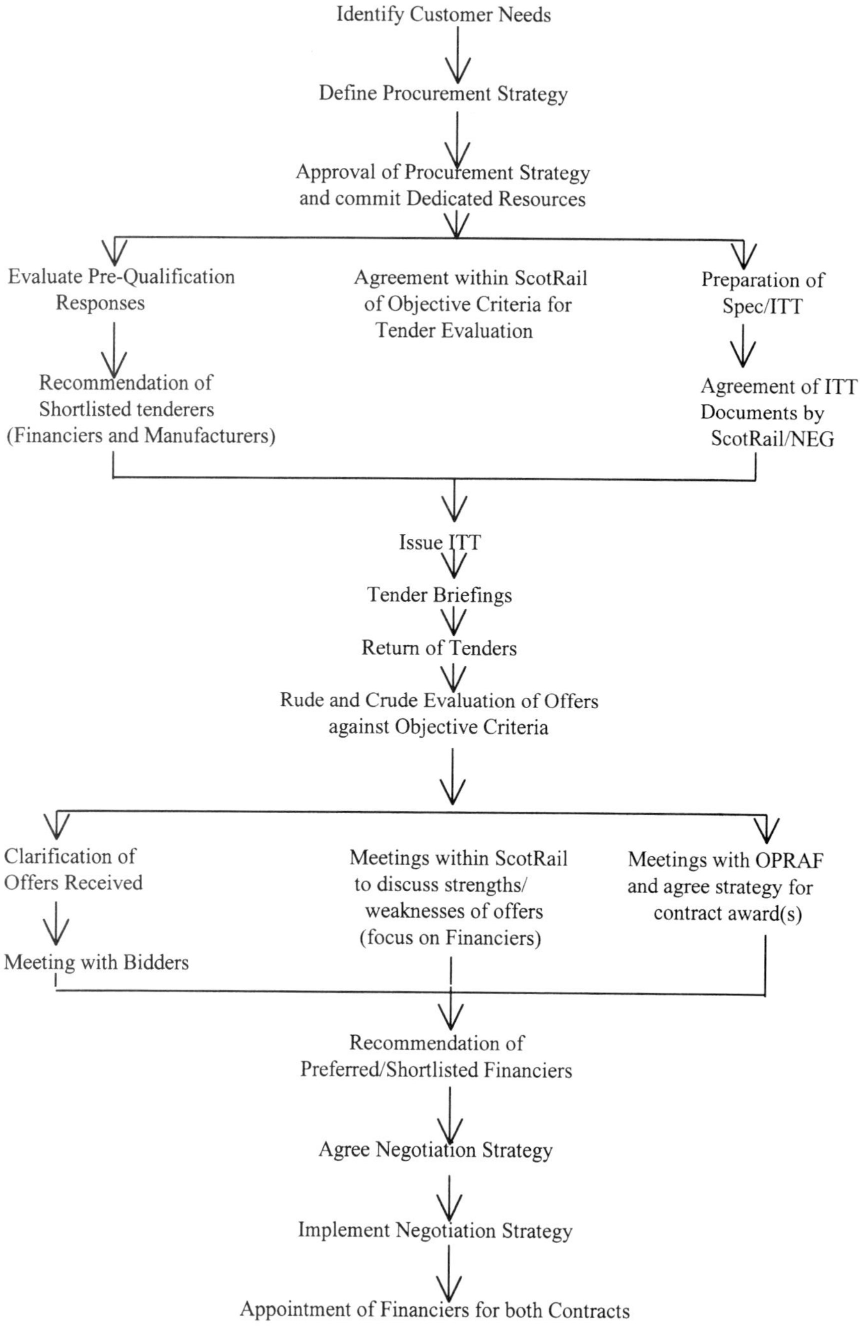


**The ScotRail Procurement Process**

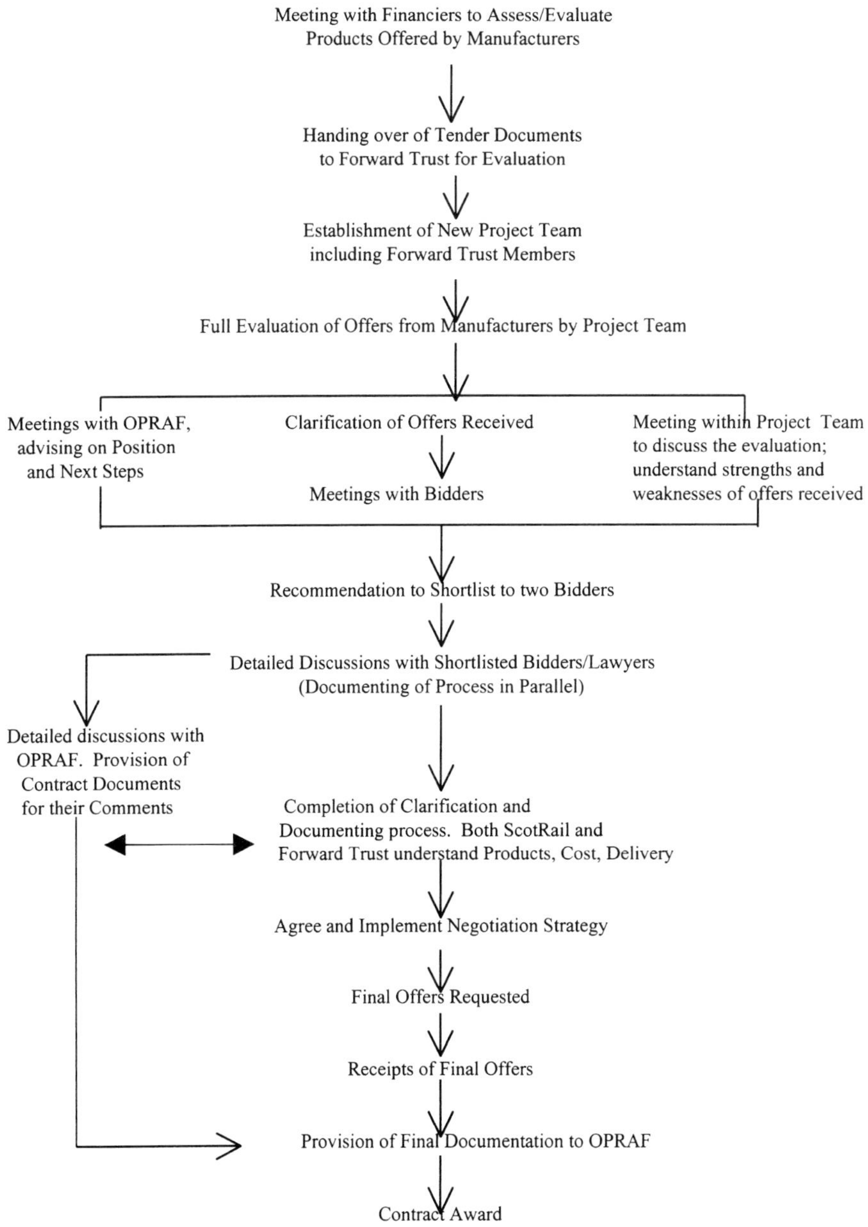


**Electric Multiple Units Only**

| | | |
|---|---|---|
| ScotRail: | Mr A Chaplin, Head of Engineering | (fax: 045 4666) |
| ScotRail: | Mr M Price, Head of Development | (fax: 045 4206) |
| Interfleet: | Mr I Mylroi, Principal Head | (fax: 05 63627) |

C552/070/98

# Making light rail affordable and attractive

**D T CATLING** BSc, MIEE, UITP, CEng, MIMechE
Urban Railway Consultant, Ely, UK
**T GRIFFIN** BTech, CEng, MIMechE
AEA Technology Rail, Derby, UK

**SYNOPSIS**

The Paper sets out a logical general sequence, allied to a flexible and imaginative approach, by which new light rail systems may be optimised and simplified and their total life costs reduced. It describes two recent major developments to reduce costs and/or increase overall patronage:-

- low floor vehicles to provide level access for mobility impaired people
- light rail vehicles sharing existing conventional rail tracks.

Examples of relatively low cost developments to increase system attractiveness are drawn from the nearly 400 worldwide light rail systems, and developments in alternative traction are summarised.

## 1 DEFINING THE TASK AND OBJECTIVES

Light Rail Transit (LRT) can be defined in terms of its role, rather than its technology, as "the application of appropriate rail technology and operating practices to an intermediate level of passenger demand, broadly in the range 5,000 to 15,000 passengers per hour (pph) in the peak direction, to provide a more cost-effective and effective service in this range than is possible with conventional bus or rail operations." (1)

LRT has less traffic and revenues than a full metro and, unlike buses or trolleybuses, has to bear the full cost of providing and maintaining its own right of way and other major equipment. It follows that, if it is to be viable, all its elements must be planned to minimise total life cycle costs and to maximise total patronage.

To achieve these objectives, and to reduce road congestion and pollution by attracting car users to a comparable and competitive alternative, an LRT system must be:-

- **safe**: in all foreseeable normal and emergency conditions, not only for passengers and staff, but for pedestrians and other users on street running sections
- **secure**: passengers must feel free from fears for their personal safety; be kept fully informed throughout their journey
- **reliable**: punctual departure; predictable journey time; maintaining booked connections
- **understandable**: ticket purchase, fares; stops, routes, information sources

• **physically accessible**: conveniently located stops; level boarding; easy interchange, especially for people with impaired mobility
• **attractive, user friendly**: pleasant to use; clean; allweather comfort; speedy, frequent, regular service patterns, generally with no need to consult a timetable
• **economical, affordable**: well planned; minimum total life costs; efficient revenue collection.

The Paper describes a systematic overall approach to this complex task, based on lessons learned from some of the nearly 400 worldwide LRT and tramway systems already built or under construction.

The planning, funding and building of even a small LRT system is an arduous and costly process. It should never be embarked on without first defining and agreeing the objectives, and then making an impartial cost/benefit analysis of all feasible alternative transit modes (2). If LRT appears to be the most cost-effective and appropriate solution, the basic assumptions and cost estimates should be recorded in an initial "Client Specification" and agreed by all interested parties, as a committed basis for the next stage.

A new LRT scheme will inevitably have to compete with other major projects for external funding. UK (3) and other (4) Governments increasingly look to the private sector, not only for funding, but to bear the short and long-term risks in planning, building and operating a system typically for 20-30 year concession periods.

The precise identification and allocation of risks between the public and private sector, the dominance of rates of return on capital on private sector thinking and actions, and the resolution of the different objectives and approaches of the private and public sectors, all combine to make the documentation and working of these hybrid agreements most complex.

It is therefore particularly necessary to ensure that the agreements do not dominate or constrain the logical development processes (set out in the next Section) to achieve an optimum final solution; or jeopardise quality, life cycle costs, operational performance, flexibility or reliability (5). Nor should they inhibit reasonable latitude to make essential changes to the Specification. Moreover, if a scheme is demonstrably well planned along the lines suggested in the Paper its chances of securing funding will be increased.

## 2 ACHIEVING AFFORDABILITY BY AN ITERATIVE PROCESS

### 2.1 Initial analysis of main capital cost elements

The above-mentioned initial capital cost estimate for the scheme should next be reviewed and, if possible, the main elements compared with a similar recent LRT scheme, to identify possible significant savings. For example, Table 1 (courtesy of South Yorkshire PTE) shows the percentage breakdown of the final *total* completed capital cost (£240M), not just the main contracts, for the initial 29km of the South Yorkshire Supertram. This steeply graded system has several major structures, a significant amount of street running, but no tunnels or shared track, and few elevated sections:-

**TABLE 1: Percentage breakdown of total final costs**

| | % | |
|---|---|---|
| CONSULTANCY AND CLIENT (Planning, getting approvals, managing) | 11 ) | |
| LAND AND OTHER RIGHTS OF WAY (Acquiring the way) | 07 ) | 28 |
| STATUTORY UNDERTAKINGS DIVERSIONS (Clearing the way) | 10 ) | |
| STRUCTURES, HIGHWAY, DEPOT (Building the tramway) | 27 | |
| PERMANENT WAY (Laying the track) | 15 | |
| ELECTRIFYING AND EQUIPPING (Completing the system) | 10 | |
| ROLLING STOCK (To carry the passengers) | 20 | |

### 2.2 Pre-construction costs

The example shows that in broad terms, over a quarter (28%) of the total cost was outwith the actual system construction, including no less than 10% for diversion of existing statutory utility services under or adjoining the onstreet alignment, before laying the track.

 

(Faced with growing pressure to reduce LRT costs, the former GEC Alsthom recently (6) offered to build complete LRT systems for a complete price including vehicles of £7M/km, but it is not clear how many of these extraneous costs would be included.)

The following could help to reduce this very heavy burden borne uniquely by LRT, compared with other surface public transport modes:-

- somehow reducing the sheer cost, time and effort required in UK to obtain the necessary legal, planning and financial powers for a new LRT scheme, in some cases nearly 20 years between initial concept and starting construction. By contrast, the French have not only managed to complete seven new or modernised LRT schemes in 10 years, but in Grenoble (7) and Strasbourg (8) have used the introduction of LRT to re-order and pedestrianise and radically improve the city centre environment. Could we learn other ways of financing and authorising new systems, e.g. local taxes to fund public transport, and devolving to regional government the legal powers to build new systems?
- the costs of diverting statutory services under the highway can be cut by reducing the amount of unsegregated street running, see Section 2.3
- a less deep track form (see Section 2.4) could also reduce the amount of utility diversion required, wherever this is a function of the minimum permitted distance between the track base and a service below, or the track form permits access to services beneath it
- in the longer term critically review the requirements for diversion and protection of utilities. These vary between different countries, and the UITP Light Rail Commission is gathering comparable data, from which it should be possible to determine if the costs in UK borne by an LRT promoter are unduly more onerous than elsewhere.

**2.3 Low cost alignment and structures**

The following can help to reduce the civil construction costs, which at over a quarter of the total (27% in our example), should be the next major target for potential savings.

2.3.1 Use or incorporate in the LRT system one or more suitable existing railway corridors in or around a city (a most valuable and irreplaceable transit asset) to achieve the following major cost reductions:-

- fewer planning risks in acquiring new land and delaying construction
- less cost and disruption than street running, or avoiding other difficult works necessary to find a suitable surface alignment in a developed city
- an older segregated alignment will probably be more straight and level than one achievable with new construction. An existing rail alignment may be completely disused, (Docklands Light Railway, DLR), handed over to the new LRT system (Manchester and Croydon), or shared with existing passenger and/or freight services (Karlsruhe, Saarbrücken, USA, see Section 4).

2.3.2 Run at ground level wherever possible, with low cost stops. Elevated running is typically six times more costly, and tunnelling between eight and twelve times the cost of a surface alignment. Underground stations are many times more costly, as well as being less accessible and secure, than simple ground level stops.

2.3.3 Use LRT's grade and curving abilities to avoid costly obstructions, e.g. providing short underpasses at congested road junctions; making minor route diversions to avoid very costly obstacles, but without significantly reducing catchment.

2.3.4 Make prudent "passive provision" for possible future traffic increases, e.g. for doubling platform length to accommodate coupled light rail vehicles (LRVs); space for increasing substation capacity; laying in pointwork, particularly on the street, for future extensions; where feasible, and possibly with external funding, allow for future statutory services alongside the track.

2.3.5 Early provision of an adequate, pitted depot is essential for stabling, maintenance and cleaning, but it need not be elaborate or even purpose-built for a small initial system. Other lower cost options include:- sharing existing facilities with a city metro or mainline railway (Vienna, Paris, Saarbrücken); a combined bus and LRT facility (Utrecht); adapting an existing building (Salt Lake City). If the system expands, a fully equipped depot can be built later in an optimum location to serve the whole network, and the initial depot used in a satellite role (DLR).

## 2.4 Track costs

Segregated track can be ballasted, is cheaper, less disruptive to lay, and is more maintainable, than on-street track, which should therefore be minimised where possible. The design, laying and maintenance of street track has to satisfy the following onerous requirements (9):-

- the grooved running rails must provide an accurate, stable, quiet, hard-wearing, running surface for the LRVs, and, with externally fed systems, a leakage free return path for the traction current
- the track base must be stable, resistant to settlement and pounding by HGVs; the filling and support and top must remain level and gapless with the adjoining road surfaces
- the top surface of the rails and any filling must not be hazardous to motor vehicles or cyclists, particularly in wet conditions
- the total installed depth must be minimised to reduce excavation and filling costs and disturbance to the road base during laying; also potentially to require less diversion of buried statutory utilities (see Section 2.2)
- the whole assembly must be capable of speedy installation, require minimum maintenance, have long life, with easy refurbishing or replacement of worn rails.

Recent developments to achieve these aims can be summarised as follows:-

- retaining standard grooved rail but reducing total slot depth by having a thinner base (10)
- increasing the rate of laying new track by further automation (10)
- using a new rail profile with shallower depth, such as the Lesley LR55, Figure 1 (11)
- experimenting with pre-encapsulating rails with the resilient polymer elastometer bonding which is often used to reduce noise and current leakage, because bonding can be difficult on site, particularly in wet conditions.

## 2.5 System development by an interactive and iterative approach

All the main interactive technical parameters (vehicle size, weight, capacity, optimum performance and service frequency; line profile, track and structures; power supply and signalling) should be examined together and iteratively to achieve the optimum solution. For example, level boarding (Section 3) and track sharing (Section 4) can reduce stop times and interstation run times respectively. The shorter total journey times will require fewer LRVs and crews for the same peak frequency. This could only be otherwise achieved at increased cost by higher performance vehicles with increased power consumption and maximum current demand.

Identify any physical constraints on the route alignment which could impose design and cost penalties on the whole present and future vehicle fleet (e.g. LRV width or length, minimum curvature or maximum gradient), and estimate the cost of eliminating them. The cost savings from using simpler and existing standard vehicles might justify spending more on the infrastructure to eliminate a pinch point.

Because of the small self-contained scale of most new LRT systems it is possible to follow this process more manageably and effectively than for say a full new or extended metro. Use a "bottom up" rather than "top down" approach, and take full advantage of the less onerous design requirements for LRVs to ensure that only the minimum means is used to achieve the objectives listed in Section 1.

Develop technical solutions, with flexibility and imagination, learning the best and worst from other comparable systems, so it is not necessary to "re-invent the wheel" for your scheme; keep technology simple especially if this is a city's first scheme; if possible start with the easiest part of the system (e.g. Sheffield fortuitously built its segregated line first on which it resolved any technical issues, and only then took on the operational problems of street running in the City centre).

## 2.6 Electrifying and equipping

The following points are noteworthy:-

- lighter LRVs reduce power demands and costs throughout the LRT electrical supply chain
- a single tensioned overhead wire without catenary can be adequate for LRT at speeds up to 80km/h; the supporting masts can then be lighter and lower, with less deep foundations
- on-street systems can operate on "line-of-sight", and only require very basic signalling where sighting distance is limited for particular reasons (15)

## 2.7 Vehicle costs

Total fleet size must be optimised as described in Section 2.5 to determine for a specified availability the minimum number of peak service and spare LRVs; the latter may be reduced by increasing spares of bogies and possibly other major units.

Costs per LRV can be reduced by avoiding over-specification; carefully weighing the merits of partially or completely low floor LRVs (Section 3); adopting an existing design already proved in service, and possibly following on another order. For example, the new Croydon Tramlink system has ordered an existing design of 76% low floor LRV already extensively proved in service in Cologne (13), in build for Stockholm, with variants in service in Vienna and Saarbrücken. Also the vehicle builders (Bombardier) are part of the Tramlink Consortium, and will maintain the vehicles. Croydon can therefore expect troublefree vehicles, with maintenance procedures already validated by long-established operators, in addition to a presumably attractive price.

The possible initial or interim use of surplus second-hand LRVs in good condition, either on cost or availability grounds, might also be considered (Essen ex DLR; Konya, Turkey ex Cologne).

# 3 IMPROVING ACCESSIBILITY: THE LOW FLOOR CAR REVOLUTION

Over the last decade there has developed a much greater international willingness (and in the EU almost a condition of funding new LRT systems) to provide for the needs of "transport disabled" or "mobility impaired" people. These total about 15% of the adult population, if we include not only those whose mobility is limited by reason of age or physical infirmity (impaired limb movement, sight or hearing), but is also "self-imposed" by luggage, shopping, baby buggies and small children. Apart from any social or legal considerations, this total represents a sizable potential market for passengers, particularly with an aging population, and those without access to private cars. (14)

Full system access requires virtually level, gapless boarding. For all new LRT systems the UK Railway Inspectorate specifies (15) maxima for height difference (35mm) and gap (40mm) between platform and LRV door threshold for safe access by unassisted wheelchairs. This can be achieved with careful design on new fully segregated systems (DLR) by having full height platforms level with the car floor, which on classic LRVs with frame-mounted DC motors is between 850-1000mm above rail level. This is fixed by the clearances required over the motored bogies, which are usually located at the outer ends of a two element articulated vehicle, with the central articular bogie non-motored.

However, level boarding is much more difficult to achieve on unsegregated street running systems, where the provision of numerous full height platforms could constitute an unsightly barrier and safety hazard to other road users. To meet this challenge Geneva in the mid 1980s successfully introduced a radically new LRV concept, but wisely based on conventional technology, with a 480mm floor height and single step at the doorways, and over about two thirds of the total LRV length. The principle was developed further in 1987 for Grenoble (7). The new "standard" French LRV design (later continued in Rouen and Paris), achieved a floor height at the doorways of only 345mm. This made level boarding possible from a safe and acceptable onstreet platform. (Figure 2)

Subsequently, vehicle development has followed two separate paths:-

1 refining the partially low floor concept, but retaining current bogie technology, e.g. by using the new, smaller AC traction motors to lower the high floor to around 580mm on the Croydon LRVs (13), with one, instead of two, steps between the high and low floor.

2 achieving a low floor over the whole passenger length of the car. This required completely different drives to the conventional DC frame-mounted motor, using small individual hub axle AC motors (Strasbourg, Brussels); motors mounted on the underframe (Bremen) or in the body side (Vienna) driving through cardan shafts and gear trains. There was an unrealistic "explosion" in only 10 years of a whole range of widely different low floor designs, given the relatively small total potential market.

While there were benefits from the parallel introduction of AC motors and variable voltage/frequency (IGBT) controls, many of the mechanical designs were complicated, costly and heavier; some were noisy both on straight and curved track, or caused vibration, rode badly or were more susceptible to track imperfections (16,18).

Of necessity, after this excess of technology-driven innovation, a more realistic and logical approach is being adopted. With a rationalisation in the LRV industry, the remaining major manufacturers are concentrating on modular designs to reduce costs but still meet users' reasonable particular requirements (e.g. vehicle length, width and track gauge), capable of being made up from a range of basic modules without incurring massive tool costs from small quantities of each variant. (19) There are also indications (17) of a welcome period of consolidation in the main traction packages after the major developments in AC traction motor drives.

Another example of potential significant cost reduction is the UK "TRAM" project. This interesting concept has applied automotive principles and components to the development and construction of a new low cost, partially low floor, high performance lightweight prototype LRV, now on proving trials in Blackpool (20).

The authors are firm advocates of providing full access for the mobility impaired, but stress the need to assess carefully the additional capital, running and maintenance costs and likely technical risks of novel designs with low floors over the whole interior length, compared with more conventional designs with low floors at the doorways and a high proportion of the length. It should also be remembered that:-

- full accessibility is only achieved if the same attention is paid to all parts of the system, including safe movement within vehicles, access to stations, (particularly elevated or in tunnel), and at interchanges
- appropriate national standards (21) must be met to cater fully for the needs of those with impaired vision or hearing, as well as people with limited movement
- a transport authority is responsible for the safety of *all* categories of passengers accepted on its system, at *all* stages of their journey, and under *all* normal and emergency situations (e.g. safe evacuation in tunnel or on a viaduct).

## 4 TRACK SHARING

The concept of track sharing, i.e. the operation of LRVs over railway tracks which are also used by conventional trains, has taken off during the 1990s, and can contribute significantly to reducing capital costs and making LRT more attractive.

Pioneering work in Karlsruhe, Germany has proved very successful (22). A new LRT service was created in 1992 between Bretten and the city centre by linking the existing tramway and railway networks by a short connection, building new simple LRT stops on the railway and buying new "dual mode" vehicles which can operate on either system. The principle has been extended to a further 9 routes, changing the character of the Karlsruhe system from a largely urban tramway into a regional network (23) with some routes 50km long. The expansion has been extremely effective in attracting more passengers, partly through the integration of different modes achieved by track sharing.

The principal technical requirements to achieve track sharing are:-

- compatible wheel and rail profiles on plain track and crossings; track gauge (although mixed gauge is possible)
- compatible civil engineering clearances for physically dissimilar vehicles
- safety systems which reduce collision risk between structurally incompatible vehicles
- achieving accessible LRT platform and floor height standards within the constraints imposed by also running wider and possibly faster conventional trains
- compatible traction systems - which may require, as in Karlsruhe, dual voltage LRVs.

This is a complex issue, which has already been the subject of several papers, (24,25) but suffice it to say that these and other technical and safety requirements can be met in many countries by various means. Usually the solutions will differ from place to place in the same country depending on local circumstances.

Reducing collision risk can be achieved by introducing trainstops or Automatic Train Protection and operating rules to ensure safe separation of dissimilar vehicles.

In the USA track sharing has been achieved by "time segregation", i.e. LRVs use the system by day and freight trains by night. Interestingly, this approach is now being proposed to introduce new regional rail passenger services with lightweight diesel railcars on little used freight railways, which will be reclassified as "tramways"!

In Germany, other shared systems have been introduced in Saarbrücken (26), crossing the border into France at Sarreguemines. Kassel has developed various ingenious solutions to introduce low floor LRVs on shared track (27), including interlacing track at shared stations (Figure 3). In Zwickau diesel cars run on tram tracks into the city.

In France (28), there are plans for LRT sharing with freight lines (Nantes, Orléans, Troyes) and upgrading existing passenger lines (Bordeaux, Mulhouse).

In the UK, Tyne and Wear is introducing the idea on its light metro extension to Sunderland, although Nottingham pioneered the principle, but may now go for parallel running. Schemes for Bristol, Cardiff, South Hampshire, the Medway towns and Midland Metro Line 3 are based on shared track. It has also been studied in detail for Strathclyde where an extensive system is possible.

Other European countries considering sharing include Austria, Italy, Luxembourg, Netherlands, Norway, Slovenia, Switzerland and Sweden. In New Zealand, Auckland has developed plans.

So within the current decade this method of achieving the desired benefit has made considerable progress.

## 5 REVENUE PROTECTION

No matter how safe, reliable and attractive an LRT system may be, it can never be financially viable unless revenue collection is assured, and fraud by passengers, staff or others is prevented. LRT's special problem as an "open" system is with access through several doors to ensure speedy boarding, on high productivity coupled LRV units up to 60m long. As with other transit modes, developments of stored fare and contactless card systems continue, but each authority has to work out the most appropriate method of operating and monitoring current ticketing systems, taking account also of the other benefits of a visible member of staff, in giving information, re-assuring passengers and deterring antisocial behaviour. The following are varied examples:-

- use of single ticket machines at stops, with onboard magnetic validators (Rouen)
- simple on-board self-service machines (Kassel, Rotterdam)
- ticket inspectors, either in plain clothes or in highly visible uniforms (Gothenburg)
- re-introducing conductors, in South Yorkshire Supertram, moving through the vehicle or in a rather special case in Amsterdam, permanently seated at the rear of single-ended cars, whom all boarding passengers must pass (29).

## 6 ATTRACTIVE "LOW" TECHNOLOGY

We need to remember that, with imagination and vision, LRT systems can also be made more attractive, safe and viable without necessarily involving advanced technology, as the following simple examples show (30):-

- in Hannover, passengers boarding at night can ask the driver to call up a taxi on the train radio to meet them at their alighting stop and take them home for a fixed fare
- in The Hague the driver can close off the rear of the tram with a special door at lightly loaded evening periods, to keep passengers in his view and reduce unruly behaviour. (The scheme was financed by increased tobacco taxes as part of a national programme for increasing security in public transport.)
- in Antwerp and elsewhere extensive cycle parking is provided at major stops to increase the catchment area and reduce door to door journey time
- LRVs have been described as the "visiting card" of a system, and bright, pleasant, well-designed vehicles can be an attraction in themselves. In Zürich, Amsterdam, Cologne and many other cities, old vehicles are refurbished, gaily painted and equipped with bar, galley and tables to use as "party trams", which can be pre-booked for social or promotional events; Leipzig in its centenary year has converted one tram to open top for summer city tours.
- in Karlsruhe the latest cars for the longest shared runs have a buffet and toilet.
- Vienna runs a circular service passing the main tourist attractions in the City centre.

## 7 ALTERNATIVE TRACTION DEVELOPMENTS

The great majority of LRT systems are powered from 750V DC overhead, with running rail return. For off-street running, higher voltages (e.g.1500V DC, Tyne and Wear) are acceptable (15), and reduce substation and overhead costs.

If in the long term overhead electrification could be eliminated, then, not only will costs be significantly reduced, but LRT will become more attractive and environmentally acceptable. This section describes some alternative developments.

### 7.1 Low level DC supply

In the UK, new segregated systems can use a shrouded conductor rail (e.g. DLR) with high platforms. To reduce costs, existing unshrouded conductor rails might be retained on new shared track routes.

Rotterdam and Amsterdam successfully operate overhead and conductor rail systems on different parts of the same line.

Ansaldo of Italy is developing a low level current collection system for onstreet use. The conductor is only live when underneath the vehicle. This concept might be applied to LRT in situations where overhead supply is considered too intrusive.

### 7.2 Onboard energy storage systems

These have a long history on tramways, but none are likely to be able to meet total power requirements unless vehicles are extremely light. For example, the Parry People Mover, which recently entered revenue service after prolonged development, uses flywheel energy storage, recharged by low voltage electrical pick-up only at stops. However, there is scope for using this type of technology over parts of a system where overhead electrification is too costly or visually unacceptable. Options include batteries and flywheels. The latter tend to be lighter, require less time to recharge, should have a longer life, but store less energy.

Current research on advanced traction batteries and fuel cells for road vehicles could, if successful and economic, be readily applied to small "ultralight" rail vehicles.

### 7.3 Diesel power

Diesel trams are also a possibility and Siemens has developed a design for manufacture. Generally, however, light rail is expected to perform better than buses in terms of environmental pollution, and diesel trams would not be acceptable. Performance would also be an issue. But the automotive industry does not stand still and cleaner diesel and gas turbine technology could eventually be of benefit to light rail.

### 7.4 Hybrid vehicles

There is scope for various forms of hybrid, especially where the the need for "clean vehicles" applies to only relatively short sections of the system, as on hybrid diesel/electric trolleybuses. Parry has developed a prototype tram where the flywheel is powered by a small engine, and the vehicle has the same performance as those recharged intermittently from an external power supply.

## 8 CONCLUSIONS AND ACKNOWLEDGEMENTS

LRT will succeed wherever it is shown objectively to be the most cost-effective and appropriate mode, provided always that:-

- each new system is developed logically, simply and attractively, to minimise capital costs, but not by compromising quality or reliability to the detriment of life cycle costs
- the temptation to re-invent the wheel each time is resisted, but full discriminating use is made of the wealth of solutions to every conceivable situation from nearly 400 worldwide LRT systems.

The Authors gratefully acknowledge the help of the Light Rail Transit Association (LRTA), Platform 5 Publishing, the Members of the UITP Light Rail Commission, and valuable comments from colleagues in the industry, but the views are the Authors' own.

 

**REFERENCES**


1 CATLING D T, Light rail concepts; ICE Conference, "Moving people in tomorrow's world", 1986

2 CATLING D T, Selecting a new rail system; UITP Revue Public Transport International, p13, 1994/4

3 TARR, R J, Private finance and light rail; ibid., p27

4 ALS, Per, European study on light rail financing; UITP Revue Public Transport International 1997/1

5 HELLEWELL, D S, What future for light rail and urban transit?, Tramways and Urban Transit, LRTA, January and February 1998

6 HAYDOCK D, Current developments in France; Light Rail Review 8, p62, Platform 5 Publishing Ltd and LRTA, May 1998.

7 TAPLIN M, Go to Grenoble!; Light Rail Review 2, p12

8 HAYDOCK D, Strasbourg's futuristic Tramway; Light Rail Review 6, p41, 1994

9 BATEMAN D, Tracks to the cities, (a light rail supplement); The Permanent Way Institution, ISBN 0 903489 02 3, 1994

10 BARBERON, M, Tramways: Les meillures voies au moindre coût, La Vie du Rail, p45, 6 May 1998

11 LESLEY L, AL-NAGEIM H, MOHAMMED F, The performance of a new low profile track system, LR55, in various loading environments; ICE Track Design Seminar, November 1997

12 BATEMAN D, Track design, construction and maintenance; ICE LRT Seminar, Infrarail 98, Manchester, October 1998

13 FOX P, Croydon Tramlink - Trams return to London; Light Rail Review 8, p49

14 CATLING D, GUILLOSSOU M, ROVERE G, STEFANOVIC G, Ease of use of light rail systems; UITP Paris Congress 1995

15 RAILWAY SAFETY PRINCIPLES AND GUIDANCE, Part 2, Section G Guidance on Tramways, HSE Books HS(G)153/8,1997

16 HONDIUS, H. R, Six years of experience with low floor trams; ICE International Seminars on Innovation in the railway system, Basel December 1996

17 HUOBER W, Improved Traction Control; ibid.

18 Low-floor LRVs move into second generation: International Railway Journal, p37, January 1998

19 TAPLIN M R, Low-floor trams: manufacturers take the initiative; Light Rail Review 7

20 LESLEY L, Applying the automotive philosophy to making affordable light rail vehicles; ISATA 23rd Conference, Stuttgart, September 1995

21 BARHAM P,OXLEY P, SHAW A G, Accessible public transport infrastructure guidelines; UK Department of Transport Mobility Unit, 1994

22 DRECHSLER G, Light railway on conventional tracks in Karlsruhe, Germany; Paper 10802, ICE, May 1996

23 KÜHN A, GRIFFIN T, Regional metros - a new role for LRT, Light Rail 96

24 GRIFFIN T, Light rail transit sharing the Railtrack system, Paper 10847, ICE, 1996

25 GRIFFIN T, Inter-operable urban rail transport; Paper C514/035/96, IMechE

26 HAYDOCK D, Saarbrücken's Light rail system opens; Light Rail Review 8, p26

27 SCHIEFELBUSCH M, Low floor trams on heavy rail tracks: the Kassel-Baunatal extension, Light Rail Review 7, p48

28 HAYDOCK D, Current developments in France; Light Rail Review 8, p60

29 FOX P, WANSBECK C J, The return of the conductor, ibid., p51

30 CATLING D T, Increasing the attractiveness of light rail systems; UITP Revue Public Transport International 1996/4

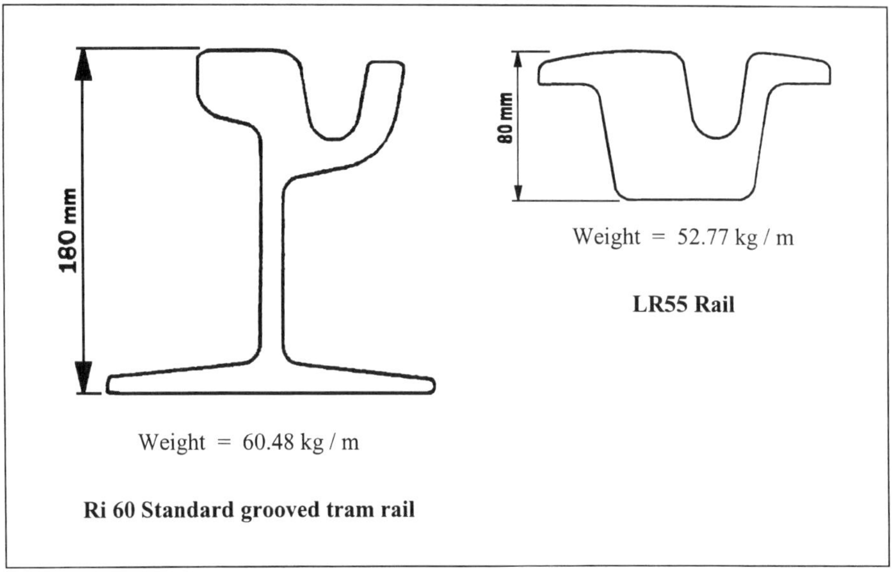


**Fig. 1 : Comparison of rail cross-sections**

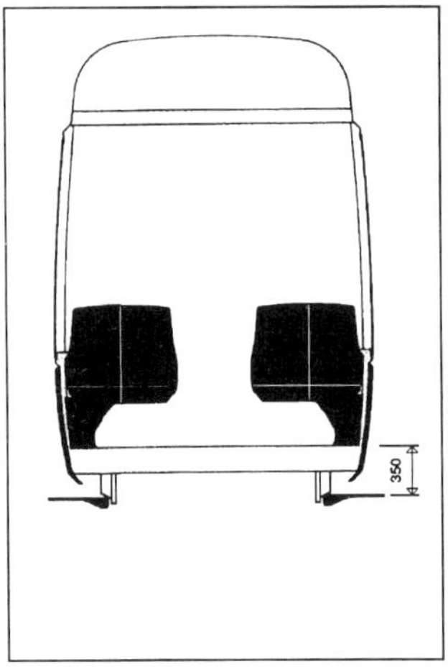


**Fig. 2 : Low floor tram cross-section**

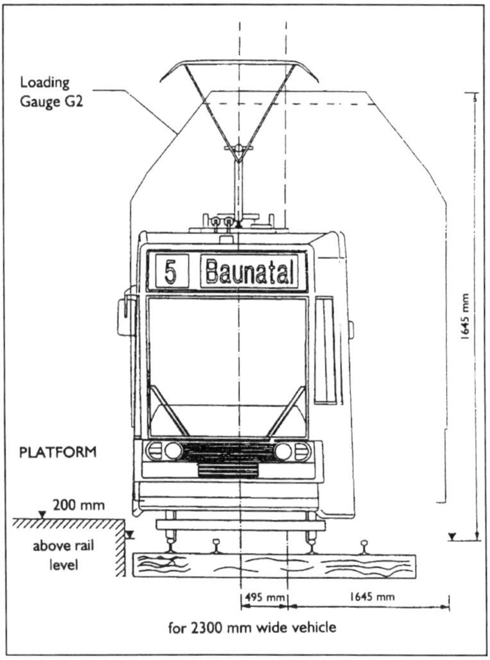


**Fig. 3 : Kassel tram on interlaced track**

C552/088/98

# Review of the environmental conditions inside an immobilized train in a tunnel

**C W POPE** BSc, MSc, PhD
Mott MacDonald Group, Croydon, UK
**D J POPE**
DERA, Chertsey, UK
**H BARROW**
Department of Engineering, University of Liverpool, UK

When trains come to rest in a tunnel, considerable quantities of heat can be transferred to the tunnel environment from the train equipment and passengers. This heat, together, with the moisture generated by the respiration of passengers can cause the temperature and relative humidity to rise to unacceptable levels. Further, in extreme cases, the carbon dioxide introduced into the interior of the train by the passengers can lead to uncomfortable or even unsafe conditions within the vehicles. The problem is dependent on the number of passengers and can be critical during the height of summer when air at high ambient temperature enters the tunnel.

A computerised prediction method has been developed and is applied to simulate the conditions that develop inside a crowded train immobilised in a tunnel on a hot summers day. The predictions are discussed in relation to comfort and safety criteria for temperature, humidity and carbon dioxide concentration.

## 1. INTRODUCTION

### 1.1 General

From time to time trains may be brought to rest in the running tunnels for extended periods due to operating difficulties or equipment failures. Under such conditions there can be a significant rise in the temperature, humidity and carbon dioxide levels inside the vehicles. Situations can potentially arise where the conditions can be a threat to the health of the passengers. The conditions are most severe in summer when air at high ambient temperature is drawn into the tunnel system from outside.

Heating of the environment within the trains is caused by the transfer of heat from the hot traction and braking equipment, the energy dissipated by on board equipment and the metabolic rate of passengers. The heat transferred to the environment by passengers is accompanied by the exhalation of water vapour from the respiratory system and perspiration through the skin. Both of these factors result in an increase in the level of humidity within the vehicles. Increases in humidity result in a lowering of the tolerance to rises in the air temperature. Heating of the environment inside the vehicles is heavily dependent on the number of passengers present and can be particularly severe if the train is heavily loaded.

An increase in the concentration of carbon dioxide arises from the air exhaled from the occupants. It is accompanied by a reduction in the level of oxygen.

Cooling and ventilation of the trains may be facilitated by:

a) the use of the tunnel ventilation system to generate a flow of air around the stationary train to increase the rate of heat transfer from the train,
b) opening of the windows to promote the flow of air into the interior to limit the increase in air temperature, humidity and carbon dioxide levels and reduce the depletion of oxygen,
c) the use of a mechanical on board ventilation system
d) an air conditioning system.

A control volume prediction method has been developed for determining the environmental conditions inside and outside a train immobilised in a tunnel. The method is applied to predict the conditions inside a non air conditioned train with opening windows brought to rest in a tunnel for one hour. The ambient conditions are taken to be those for a warm summer's day: dry bulb temperature 28°C and relative humidity 46%.

The paper comments on commonly used environmental criteria which may be applied to assess the comfort and well being of passengers on the train.

## 2. PREDICTION METHOD

In the simulation, the section of tunnel containing the train is subdivided into a series of elements corresponding to the length of each car. Within each element the conditions within the coach and the annular gap between the coach and the tunnel are computed using separate control volume analyses. The conditions within the control volumes are modelled using the unsteady form of the first law of thermodynamics. A time marching procedure using small time steps is implemented to determine the behaviour of the environmental conditions internal and external to the cars. The calculation allows the effects of a wide range of conditions to be examined. Particular features of the simulation method are as follows:

(a) a metabolic model for the passengers allowing for the effects of respiration (i.e. the production of water vapour through the exhalation of air and perspiration, generation of carbon dioxide and depletion of oxygen). The model uses a linear relationship developed by Kimura [1] for the variation of the behaviour of the sensible heat with the temperature of the environment.

(b) modelling of the flow in the annulus with heat transfer to and from the train and the tunnel walls
(c) a transient heat transfer model for the rock mantle surrounding the tunnel
(d) a composite heat transfer model for the train body allowing for an outer skin, a layer of insulation and an internal panel with a correction factor for the effects of window glass
(e) the capability to simulate the cooling of traction and braking equipment
(f) inclusion of the energy inputs from auxiliary equipment and lighting
(g) the ability to model forced vehicle pressure ventilation and inward and outward flow through open windows or apertures.

The modelling approach permits the interactive effect of successive vehicles to be simulated.

The method is similar to an approach developed by Dunbar et al [2] but embodies a composite wall model for the train sides.

## 3. ENVIRONMENTAL CRITERIA

### 3.1 Thermal Criteria

In underground metro systems, it is common to make assessments of the acceptability of the environments using dry bulb temperatures or a parameter known as the Relative Warmth Index (RWI). Frequent use is also made of the Effective Temperature.

#### *3.1.1 Dry Bulb Temperature*

The Station Planning Guidelines [3] developed by London Underground Limited stipulate that the dry bulb air temperature should not exceed 35°C. In itself, however, the specification of a limiting dry bulb temperature is insufficient since the comfort and well being of passengers depends also on the relative humidity.

#### *3.1.2 Relative Warmth Index*

The RWI was specifically developed for thermal environment assessment in metro systems and is based on the thermodynamic relationship governing the human heat balance [4]. As such it is a very comprehensive comfort assessment parameter and depends on the:

- metabolic rate
- effect of clothing insulation
- insulation effect of the boundary layer
- dry bulb air temperature
- radiant heat.

The reader is referred to Ref 4 for details of the equations and a detailed discourse on the basis and characteristics of the index. Limiting indices are as follows:

**Table 1: Description of Relative Warmth Index values.**

| Comfort Classification | Relative Warmth Index |
|---|---|
| Extremely uncomfortable | 0.5 |
| Warm | 0.25 |
| Slightly warm | 0.15 |
| Comfortable | 0.08 |
| Slightly cool | 0 |

In relation to riding on a metro, passengers would be dressed with the external ambient conditions in mind. For instance, on a cold winter's day they would be wearing overcoats giving a relatively high clothing insulation factor. On a hot summer's day light clothing would be worn giving a much lower insulation factor.

### *3.1.3 Effective Temperature*

The Effective Temperature (ET) [5] is the temperature of an environment at 50% relative humidity that results in the same total heat loss from the skin as in the actual environment. It is a function of the operative temperature, skin wetness, clothing insulation and air movement.

The operative temperature is a weighted average of the temperature of the air in which the subject is located and the temperature of any surrounding enclosure that radiates heat. In the case of the interior of a train in a tunnel the internal panelling is close to the air temperature of the inside of the train. As a result, the operative temperature is approximately equal to the air temperature of the train interior.

Two separate conditions which have different dry bulb temperatures and relative humidities but the same Effective Temperature result in the same physiological responses to the conditions.

The Effective Temperature thresholds in terms of comfort and the corresponding physiological effects for slightly cool to unacceptably hot environments are given in the table below:

**Table 2: Description of Effective Temperature values**

| Effective Temperature (°C) | Sensation | Physiological Effect |
|---|---|---|
| 41.5 | Limited tolerance | Body heating – failure of regulation |
| 40 | Very hot, very uncomfortable | Increased stress caused by sweating and blood flow |
| 35 | Hot | |
| 30 | Slightly warm | Normal regulation by sweating and vascular change |
| 25 | Neutral, comfortable | Regulation by vascular change |
| 20 | Slightly cool | |

Heat stroke can occur at Effective Temperatures above 35°C. An Effective Temperature of 41.5°C is intolerable and constitutes a dangerous condition. It is inadvisable that Effective Temperatures should be permitted to exceed 35°C for prolonged periods.

A detailed treatment of the Effective Temperature with the methodology for its derivation is given in Ref 5.

### *3.1.4 Comparison between comfort indices*

The LUL Station Planning Guidelines dry bulb temperature limit and constant RWI and ET lines corresponding to the principal comfort thresholds are compared on a standard psychrometric chart [6] in Figure 1. The RWI lines are constructed for a clothing insulation factor of 0.2 clo (appropriate to light summer clothing) and no solar radiation to be applicable to conditions inside the train. The ET lines are appropriate to a general indoor environment with people in a sedentary state. The lines were taken from Ref 4 and are broadly applicable to the environment inside a vehicle loaded with passengers.

Figure 1 also shows the comfort zone. This lies between the Effective Temperature lines of 20°C and 26.1°C. Good correlation exists between the RWI and Effective Temperature lines with respect to the various comfort thresholds. For instance, the RWI line of 0.25 and the ET line of 30°C correspond to the condition "warm" and lie close to each other.

The chart underlines the deficiency of merely specifying a dry bulb temperature limit alone. Above a relative humidity of 83%, the conditions become intolerable at this temperature.

## 3.2 Carbon Dioxide and Oxygen

### *3.2.1 Carbon Dioxide*

The concentration of carbon dioxide in ambient air varies between 0.03 and 0.06%. The ASHRAE Fundamentals Handbook gives 0.25% as an acceptable limit. Long (eight hour exposure period) and short (ten minute exposure period) term limits of 0.5% and 1.5% respectively are specified for the gas by the British Health and Safety Executive. On this basis, it is suggested that 0.5% is adopted as a comfort limit. Ref 2 proposes 1.5% as a safety limit requiring train evacuation.

### *3.2.2 Oxygen*

The normal concentration of oxygen in air is between 20 and 21%. The comfort of the individual, however, is not seriously affected unless the level falls below 16%. The British Health and Safety Executive stipulate that the concentration should be no less than 18%.

## 4. DATA INPUT

The modelling is conducted for a six car train in a double track tunnel using typical items of data for the purposes of demonstration. The principle items of data are as follows:

### 4.1 Atmospheric Conditions

Dry bulb temperature: 28°C, Relative humidity: 46%

### 4.2 Tunnel Geometry

Cross Sectional Area: 40.81m², Perimeter: 26.55 m, Friction factor: 0.01 (annulus)

### 4.3 Train Data

Cross Sectional Area: 7.9 m², Perimeter: 11.7 m, Friction factor: 0.009
Length of each car: 17 m, Open window area: 2.85 m² per car
No of people per car: 144 (i.e. 864 people per train)
Heat output from lighting per vehicle: 460 W

- Heat transfer through train body

Train body sides, ends and roof modelled as three component composite walls comprising:

(a) internal laminate panel :
thickness: 0.003 m, thermal conductivity: 0.357 W/m.K, density: 1600 kg/m³
specific heat: 1470 J/kg.K

(b) air gap and insulated layer :
modelled as single layer of fibre glass by scaling thickness to allow for air gap
effective thickness: 0.0535 m, thermal conductivity: 0.032 W/m.K, density: 24 kg/m³
specific heat: 700 J/kg.K

(c ) steel outer layer :
thickness: 0.0015 m, thermal conductivity: 54 W/m.K, density: 7800 kg/m³
specific heat: 465 J/kg.K

Effective surface area of interior of train: 170m²
Heat transfer coefficient for the interior of the train: 1 W/m².K
Heat transfer coefficient for the exterior of the train deduced from Reynolds analogy
Heat transfer characteristics for traction and braking equipment

(a) Traction equipment: cooling rate: 0.001 K/K.s
(b) Braking equipment: cooling rate: 0.0011 K/K.s

(based on concept of Newtonian cooling)

### 4.4 Tunnel Lining Heat Transfer

thermal conductivity: 1.73 W/m.K, density: 2410 kg/m³, specific heat: 840 J/kg.K
sink temperature: 11°C.

### 4.5 External ventilation flow

Flow in unobstructed tunnel: 0.5m/s (typical general air flow produced in tunnels by meteorological effects). The flow direction is taken from vehicle 1 to 6.

### 4.6 Passengers - Physiological Effects

(a) Metabolic Effects

Metabolic rate: $Q_t$=125 W

A significant proportion of the perspiration from the body enters the clothing. This reduces the amount of moisture passing into the air inside the vehicle and has an important effect on restricting the increase in the humidity of the air. The effect of the clothing in absorbing moisture has been estimated from an experiment to examine the change in thermal environment within a small compartment containing a clothed seated person [7]. Calculation showed a much more rapid rise in the dew point temperature than that obtained from measurements with a hygrometer. This indicated that the clothing of the subject was absorbing a significant proportion of the moisture from the skin. Analysis of the rise in the dew point temperature showed that following an initial period of stabilisation only 27% of the moisture generated by the body was entering the air in the compartment. This factor has, therefore, been applied in the analysis to allow for the reduction of moisture entering the interior of the vehicle due to the absorption effect of the clothing of the passengers.

(b) Carbon Dioxide Production

Rate of production of carbon dioxide per person: $4.72 \times 10^{-6}$ m³/s (Ref 8).

(c) Absorption of Oxygen

Rate of absorption of oxygen per person: $5.9 \times 10^{-6}$ m³/s (Ref 8).

## 5. RESULTS OF THE SIMULATION

### 5.1 Dry bulb temperatures in annulus

Figure 2 shows the air temperature in the annulus to rise initially. This is due largely to the cooling of the hot brakes and traction units. As the brakes and traction units cool a temperature gradient is established along the train with the highest temperature at the rear. The temperature gradient is a consequence of heat transfer to the tunnel wall.

### 5.2 Dry bulb temperatures in the train

Initially, Figure 3 shows peaks in the dry bulb temperatures to occur inside the third to the sixth vehicles. These are principally the result of air in the annulus heated by the cooling of the traction and braking equipment passing into the vehicles through the open windows. The temperature peaks increase in magnitude from the rear (vehicle 1) to the leading vehicle (vehicle 1) due to the cumulative effect of the heating along the train. The heating effect from the cooling traction and braking equipment is not sufficient to produce initial temperature peaks in the first and second vehicles. In these vehicles, the rise in temperature is more progressive and is more heavily influenced by the heating effect of the passengers and lighting. The magnitude of the rise in temperature is reduced by the effect of heat transfer to the tunnel.

Following the initial peaks, there are falls in temperature in the third to the sixth vehicles due to heat transfer to the tunnel wall. Subsequent to the falls, a rise in temperature occurs as a result of the heating effect of the passengers and the lighting after the traction and braking equipment has substantially cooled down. At 1 hour, a fall in temperature of 2.5°C occurs between vehicle 6 and vehicle 1 as a result of heat transfer to the tunnel. It is interesting to note that the effect of the heat transfer to the tunnel more than counters the cumulative heating effects of heat transfer from the traction and braking equipment and the passengers and lighting. A peak temperature of just under 32.5°C is developed in vehicle 1 at the end of the prediction.

### 5.3 Relative humidity in the train

Figure 4 shows a general increase in the relative humidity levels in the vehicles with time. The highest levels occur in vehicle 6 and assumes a maximum level of 85% after 1 hour. The cumulative increase along the train is the result of humid air passing from vehicle to vehicle through the windows. Initially it will be noticed that there is a small change of gradient after 100s. This is due to the heating of the inlet air to the vehicles by the traction and braking equipment causing a fall in the dew point temperature running from vehicle 1 to 6.

### 5.4 Internal carbon dioxide concentration

The carbon dioxide level rises progressively to concentrations that stabilise at about 0.6%. The concentration increases slightly from vehicle 1 to vehicle 6.

### 5.5 Internal oxygen concentration

The oxygen level progressively falls to concentrations in the region of 19.9%. In line with expectations, the concentration decreases slightly from vehicle 1 to 6.

### 5.6 RWI inside the vehicles

The most uncomfortable conditions occur in vehicle 1. Here the dry bulb temperature and the relative humidity rise to 32.7°C and 74% after 1 hour. This corresponds to an RWI of 0.44.

### 5.7 Comparison with ambient conditions and comfort indices

The point pertaining to the condition in the sixth vehicle after 1 hour is shown on the psychrometric chart presented in Figure 1. This condition corresponds to an Effective Temperature of just under 36°C (hot).

For comparative purposes points are given on the chart which show the hottest conditions that can arise outdoors in a number of hot cities around the world.

## 6. CONCLUSIONS

A simulation has been undertaken for a heavily loaded stationary six car train brought to rest in a section of running tunnel for an extended period with:

- an ambient temperature of 28°C
- the train windows open
- a small typical background air flow of 0.5m/s caused by meteorological effects.

It is shown that:

The highest RWI occurs inside the first car and assumes a value of 0.44 after one hour.

Carbon dioxide and oxygen concentrations of approximately 0.6% and 19.5% prevail in the vehicles after half an hour. These levels lie well within safety limits.

For the situation considered, the conditions developed within the vehicles are approaching uncomfortable. The main factor which contributes to the development of an uncomfortable environment is the moisture generated by the metabolism of the passengers. An increase in the relative humidity is likely to result in a requirement to evacuate the vehicles.

With suitable adaptation of the program, wet tunnel walls and train surfaces can be modelled.

## 7. ACKNOWLEDGEMENT

The authors thank the Directors of Mott MacDonald for permission to publish this paper.

## 8. REFERENCES

1. Kimura, K. I. "Scientific basis of air conditioning", App. Sc. Pub. Ltd, London, 1977.

2. Dunbar, I.H., Ketchell, N. and Kissane, M.P. “Passenger Safety on Immobilised Trains”, Report on Contract No. GT 5874, London Underground Limited, AEA Safety and Reliability, June 1990

3. London Underground Limited, Station Planning Guidelines, 55 Broadway, London, SW1H 0BD, March 1993.

4. Associated Engineers, "Subway Environment Handbook Vol 1 - Principles and Application", Urban Mass Transport Administration, Technical Report UMTA-DE-06-0010, Sup Docs. US Printing Office, Washington DC, Stock No 050-014-00005-3, 1975.

5. ASHRAE Handbook, Fundamentals, American Society of Heating, Refrigerating and Air Conditioning Engineers, Atlanta, USA, 1989 Edition.

6. Chartered Institute of Building Services Engineers (CIBSE), "Psychrometric Chart - For dry bulb temperatures 10°C to 60°C", 222 Balham High Road, London, SW12 9BS, 1970.

7. Barrow, H. and Pope, C. W. "Theoretical and experimental investigation of changing atmospheric conditions inside an occupied confined space with particular reference to a stationary railway vehicle in a tunnel", Ninth International Symposium on the Aerodynamics and Ventilation of Vehicle Tunnels, Organised by BHR Group, Cranfield, UK and held in Aosta, Italy, 6-8 Oct 1997.

**8.** Jones, W. P. "Air Conditioning Engineering", Edward Arnold, 1973.

## 9. ABBREVIATIONS

The following abbreviations are used on Figs 2-4.

tfl - Average flow dry bulb temperature external to coach (°C)

tb - Average dry bulb temperature inside coach (°C)

rh - Relative humidity (fraction - 1 indicates saturated conditions)

The number following the abbreviation is the number of the vehicle. Vehicle 1 is at the inlet end of the tunnel.

**Figure 1 Environmental indicies and predicted conditions**

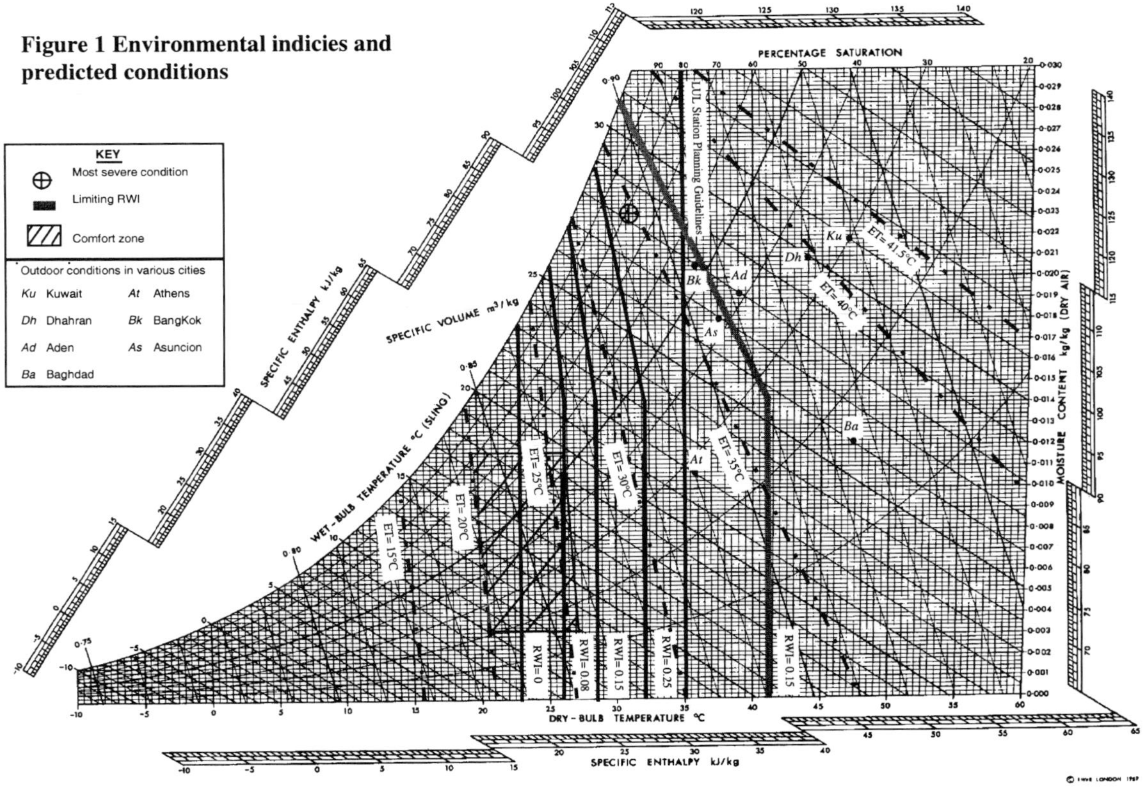

Figure 2 Variation of external temperature (deg C) with time (s)

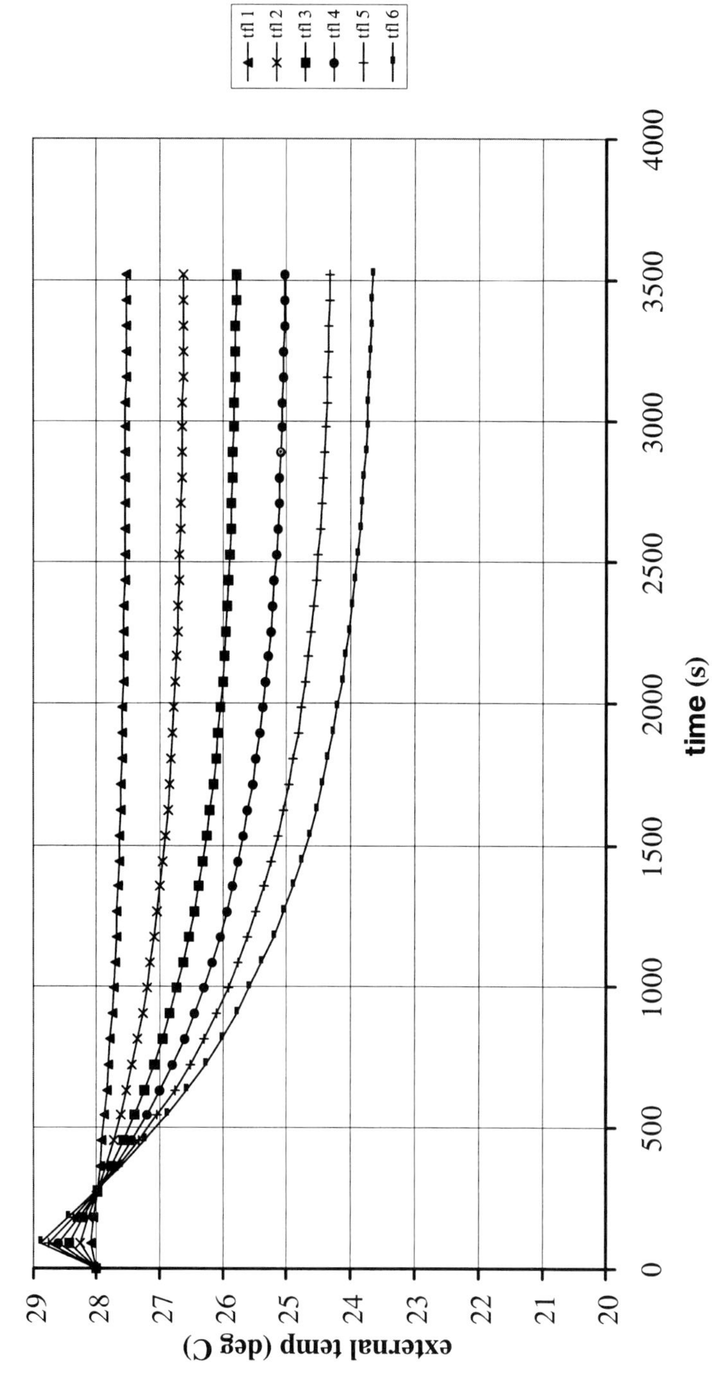

Figure 3 Variation of internal temperature (deg C) with time (s)

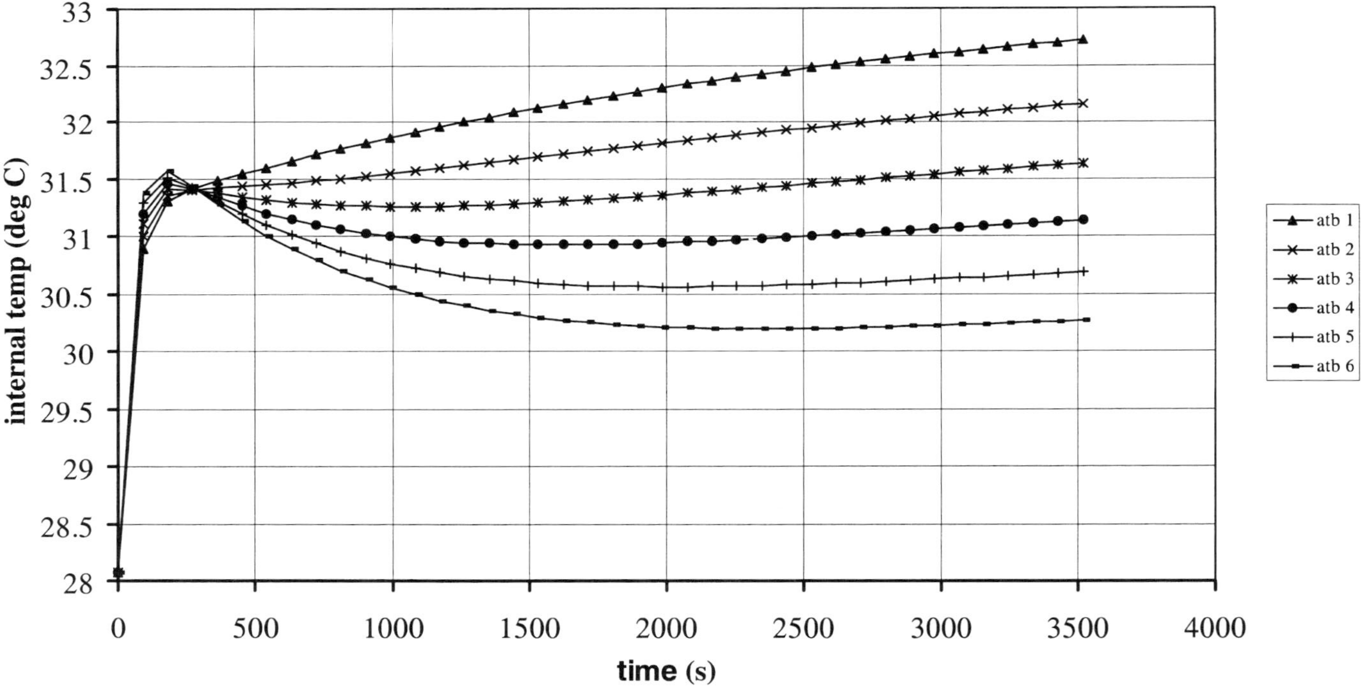

Figure 4 Variation of relative humidity with time (s)

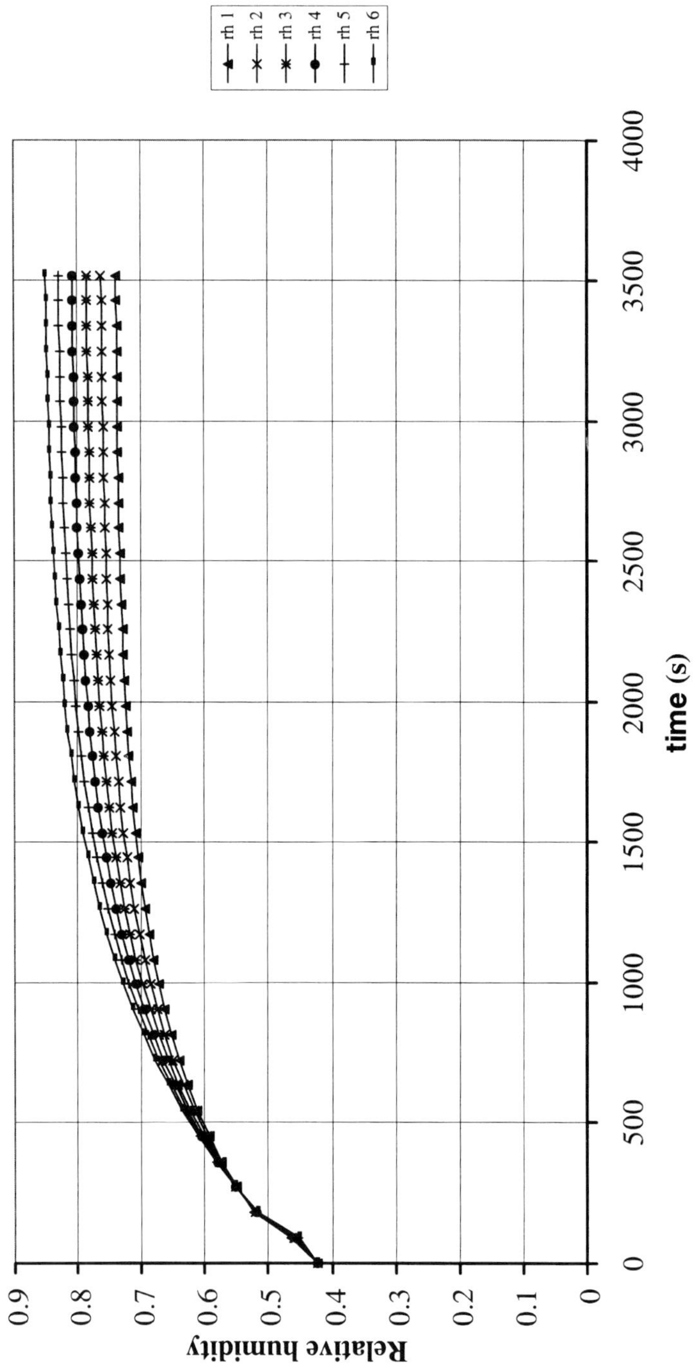

C552/038/98

# Steer-by-wire – a railway technology for the future?

**R M GOODALL** MA, PhD, FIMechE
Department of Electronic and Electrical Engineering, Loughborough University, UK

## 1. INTRODUCTION

Over the last two decades there have been varying levels of interest in the use of active suspensions for railway vehicles (1-3). Significant use however has only happened during the 1990s with the advent of tilting trains in service operation, and tilt is now approaching the stage of being accepted as standard for many high speed passenger trains and increasingly for regional/cross-country services (4). This acceptance of tilt has paved the way for other uses of active control within the vehicles' suspensions, and it seems likely that more general use of active secondary suspensions will progressively increase. However, although there has been a small amount of experimental work, active primary suspensions remain a largely theoretical possibility.

This paper presents a speculative look at active primary or "steer-by-wire" solutions for rail vehicles. It describes the basic principles, reviews the theoretical and experimental state-of-the-art, and highlights the key technological challenges which need to be faced.

## 2. BACKGROUND

### 2.1 Wheelset mechanics

The solid-axle wheelset has been a fundamental component for rail vehicles since the earliest days. Originally there was no coning or profiling on the wheels, but the benefits of coned instead of cylindrical wheels were clearly identified in the 1830s and their use has become standard practice ever since (5). Despite this, it wasn't until the 1960s that the science of vehicle dynamics using coned wheelsets was comprehensively understood (6), and the analytical techniques which emerged enabled substantial refinements and innovations to be made. However, the design trade-off between ensuring dynamic stability and providing good curving performance remains, because only stiffness can stabilise the wheelset(s) and this inevitably leads to less than perfect curving action.

Throughout the history of railways people have also investigated wheelsets in which the two wheels are independently rotating. This removes the intrinsic mechanical steering action of the solid-axle wheelset, but is much more straightforward to stabilise, and significant engineering development has been applied to vehicles having independently-rotating wheels (7).

## 2.2 The rationale for active control

It is obvious to any practising engineer that the idea of replacing a high-developed well-proven purely-mechanical device such as the railway wheelset with a new unproven system relying upon an arrangement of sensors, control processors and actuators should not be considered lightly. So what are the reasons for considering active control of the primary suspension?

The rationale is threefold. Firstly, as mentioned above, the trade-off between curving and stability of conventional wheelsets is still a critical problem in vehicle design, and the incorporation of active control provides a design freedom which is not available in a practical mechanical suspension. This is because a mechanical component such as a spring or damper can only produce a force which is dependent upon the displacement or velocity applied between its ends, whereas the force produced by an actuator can be dependent upon a measured variable from elsewhere in the mechanical system, and independent of the motion of the actuator.

Secondly, the typical mechanical arrangement with a body, two bogies and four axles has evolved to help overcome the stability/curving trade-off, but it is mechanically complex; if the wheelsets are actively controlled, perhaps it is possible to revert to a much simpler and lighter (and cheaper?) mechanical arrangement.

Thirdly there are a number of examples of innovations in other industrial sectors where a mechanical engineering solution has been replaced by an electronic system. Once the challenges have been faced then significant benefits, possibly unseen at the outset, have been realised. Good examples are the "fly-by-wire" aircraft and the CD player, the latter being a particularly significant because it required a fundamental change to the basic recording medium and method, and must have seemed a daunting prospect in the early days of development!

## 2.3 Actively-controlled vehicle configurations

Before thinking in more detail about how the wheelset might be controlled, it is useful to follow-up the comments made in the previous sub-section about possibilities for the basic mechanical arrangement of the vehicle to which active control is to be applied, because this will inevitably have an important effect upon the formulation and performance of the control laws. One option of course is the conventional bogie vehicle, in which ordinarily the bogie and its wheelsets are substantially decoupled from the vehicle body in a dynamic sense. This leads to a simpler control design because each bogie can be considered separately (although the possible value of control linkages between the two bogies should not be ignored). In addition, the well-established expertise in bogie design can provide valuable insight into the formulation of the control laws.

The natural extension to this is to consider two-axle vehicles, which are potentially of great interest for reducing weight and mechanical complexity. The essential dynamic complexity of control will be similar to that for the bogie of a conventional vehicle, although it is a more profound problem to achieve the required performance, for example because of the increased yaw angles required for the wheelsets during curving.

A third configuration is to incorporate articulation, the "ultimate" being single-axle vehicles of a form similar to the new S-Tog trains for Copenhagen (8). This configuration is dynamically complex, and it is possible that the most effective form of control will involve a complex "train-wide" control law in which a centralised controller co-ordinates the forces applied to all the wheelsets, possibly controlling riding quality as well as curving and stability.

The issue of vehicle configuration has been raised because in practice the real performance which is achievable through active control can only be predicted in the context of the actual vehicle arrangement. However the principles of active primary control are still emerging, and so it is useful to consider what control approaches might be useful at a more local level, i.e. the wheelsets, and this is what is covered in the next section.

## 3. CONTROL POSSIBILITIES

There are three types of wheel mechanism which can be considered, each having quite distinct control options because of their particular characteristics/limitations: a solid-axle wheelset, a wheelset with an axle but having independently-rotating wheels, and wheel pairs without an axle but which are directly steered (e.g. the kind of steering mechanism found in automobiles). Control options for these are suggested separately in the following sub-sections.

### 3.1 Solid-axle wheelset

There are essentially two possibilities here. The two methods have quite different sensing and actuation requirements, and their failure modes will be significantly different.

In one approach the wheelset can be stabilised through normal passive means, with the active control used to overcome the degradation in curving ability caused by the passive stabilisation. This steering control will essentially be "low bandwidth", requiring action only at low frequencies for curving; also the action must of course be applied in order to yaw the wheelset.

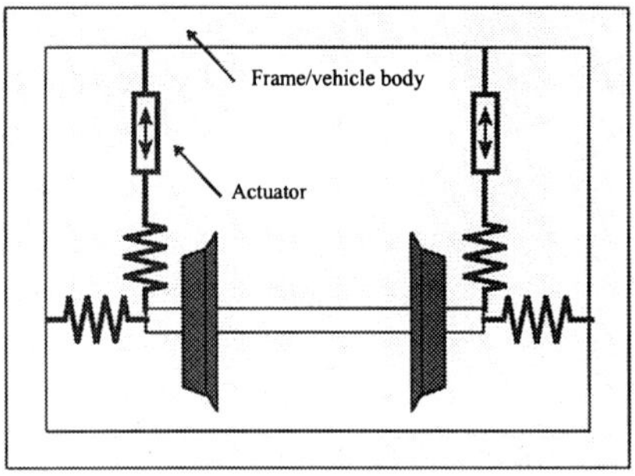

**Figure 1** Actively-steered conventional wheelset

Figure 1 shows a typical scheme in which two linear actuators are used differentially to apply the steering action via the longitudinal springs which stabilise the wheelset. The other option, not indicated in the diagram, is to apply a yaw moment directly to the wheelset, in which case the steering moment will have to be applied to overcome that generated by the longitudinal springs.

The other method is to use the intrinsic mechanical steering of the wheelset, with the control action providing active stabilisation of the wheelset dynamics in a manner which does not interfere with the wheelset's curving action. This must be high bandwidth control because it must be effective at the kinematic frequencies, and in principle lateral and/or yaw action is possible (Figure 2). For example, it can be shown that stability can be achieved either using a yaw torque proportional to lateral velocity, or a lateral force proportional to yaw velocity, neither of which will affect steady-state curving. (Note that both these can only be achieved through active control.)

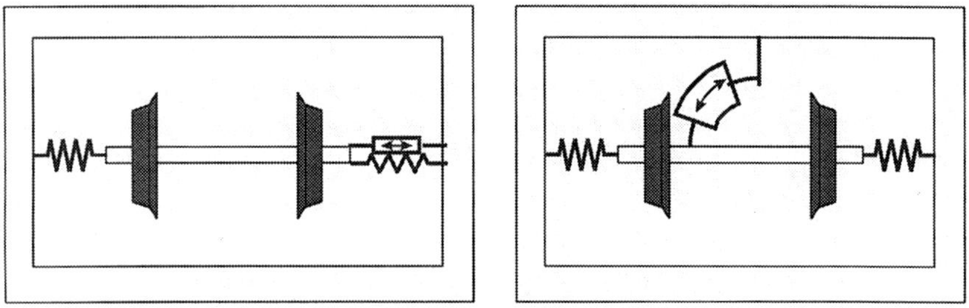
**Figure 2** Actively-stabilised conventional wheelset - lateral and yaw control

### 3.2 Wheelset with independently-rotating wheels

The natural mechanical steering action is lost when the two wheels on a wheelset are uncoupled, and so the primary purpose of the active control must be to steer the wheelset, for which various approaches are available. For example the active control can be used to align the wheelsets radially on a curve, although cant deficiency forces can only be provided by the wheel flange in this case - a small angle of attack is needed to develop lateral creep forces via the wheel tread. Alternatively, the wheelset can be steered to keep it central within the flangeway clearance (or at least off the flanges); this obviously requires some measurement of the lateral wheel/rail displacement, which is not a straightforward measurement.

It is important to realise that a wheelset with independently rotating wheels still exhibits a quasi-kinematic oscillation which must be stabilised; whether this is achieved passively, or via the active control, is another question to be addressed.

 

Control must be applied in the yaw direction; this might be via a rotary actuator or a pair of linear actuators applying yaw control directly to the wheelset. An alternative is where differential control of the traction/braking torque is applied to the two wheels (see Figure 3 for conceptual schemes). (The latter case is an interesting but non-trivial option because the torques required for wheelset control will be many orders of magnitude lower than that required for traction/braking.)

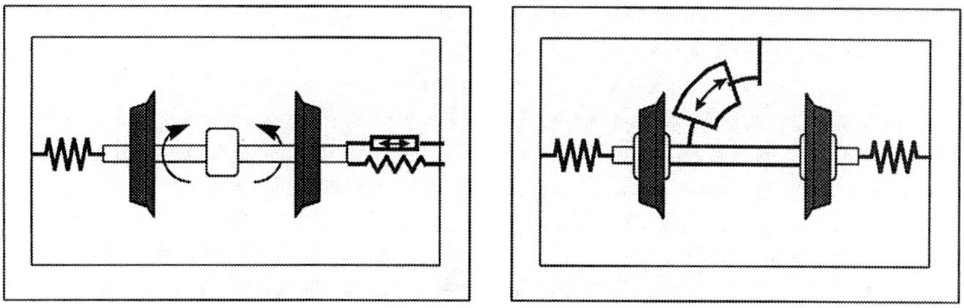

**Figure 3** Active wheelset with independent wheels - torque and yaw control

### 3.3 Directly-steered wheel pairs

This approach represents a radical departure from current railway engineering practice because the axle has been dispensed with, and replaced by a steering mechanism - Figure 4 shows a typical scheme of a wheel pair. As with the arrangement of the previous sub-section, control can either be applied by means of a steering actuator fitted to the mechanism, or via differential torques applied to the two wheels. The control strategy is likely to be based upon making the steering angle dependent upon the lateral wheel/rail displacement (9).

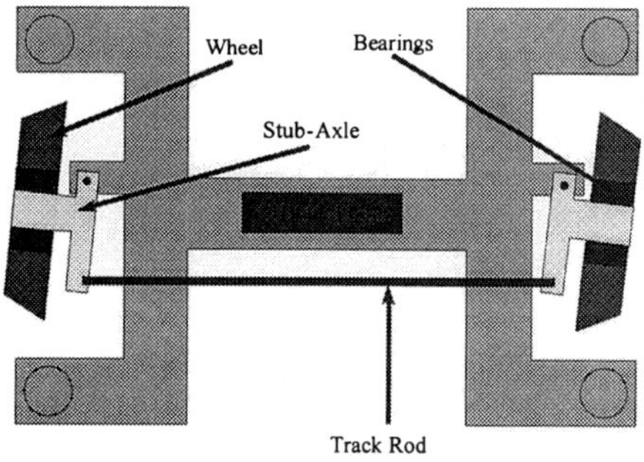


**Figure 4** Steered independent wheels

## 4. STATE-OF-THE-ART OF ACTIVE PRIMARY SUSPENSIONS

This section presents a summary of known studies which have been undertaken. The review is primarily based upon work which has involved experimental verification, but some theoretical studies which exemplify the options described in the previous section have also been included.

## 4.1 Experimental studies

One of the earliest developments was a "creep-controlled wheelset", a semi-active primary suspension using a variable electro-magnetic coupling between the two wheels in the wheelset (10). The basic idea was to shift some of the creep from the wheel contact area to the coupling, thereby diminishing the longitudinal creep. Comprehensive testing was undertaken on a roller rig with speeds of up to 500 km/h, but the work has not been continued since the 1980s.

During the 1980s a significant programme of research was undertaken by the French research organisation INRETS. One concept named ERIOM, carried out for Matra-Transport and RATP (the Paris urban network), studied an electro-magnetically steered wheelset with independent wheels. A quarter scale model was tested experimentally on a roller rig at Grenoble (11), and demonstrated that the independent wheels provide good curving behaviour, while good stability at high speed is given by a specific control law. The model was also used to measure creep forces at the wheel-rail contact (12). Further work was carried out for GEC-Alsthom, and employed hydraulic actuators to provide steering of axles with independently rotating wheels using eddy current sensors measuring the lateral displacement (13). An experimental study involving a two-axle 25 tonne freight vehicle with a wheelbase of 10m was completed; several configurations were patented but details remain confidential. Concerns about meeting failure requirements have meant that the project has not continued since that time.

Another active steering scheme using independently rotating wheels has been tested on a Rotterdam tram (14). A hydraulic steering system is controlled by signals both from a gyroscope detecting the yaw velocity of the vehicle body and from an angular displacement device between adjacent articulated bodies. A similar concept has also been implemented using purely hydraulic steering (i.e. no electronic control) based upon the angle between adjacent vehicles (8), designed by Linke-Hofmann-Busche for the Copenhagen S-Trains.

A theoretical and experimental assessment of semi-passive control of bogies with unsymmetric suspensions is another area which has been studied, and this work included testing of trial bogies by the Central Japan Railway Company (15). The basic idea uses conventional bogies with solid-axle wheelsets, in which stiffnesses are switched depending upon the direction of motion to make the longitudinal or yaw restraint of the leading wheelset much softer than that of the trailing wheelset. An alternative uses switched damping. The overall conclusion of the study was that dramatic reductions in the lateral forces on the wheels could be obtained during curve negotiation.

A 1/5 scale roller rig which has been designed for active primary suspension research has been developed at the German research establishment DLR-Oberpfaffenhofen (16). The wheelset structure is being equipped with electric actuators to give a controlled differential torque to the wheelset by which a range of active or semi-active couplings can be implemented. One configuration demonstrates very effectively the use of differential torque to provide centralisation of a wheelset with independently rotating wheels.

A recent study by DB, in collaboration with Aachen Polytechnic, has also applied differential wheel torque (17), this time to a full size two-axle vehicle. Details of the control strategy are not given, but it is likely that this is the most advanced full scale active primary suspension which has been developed.

### 4.2 Theoretical studies

There have been a relatively small number of conceptual and theoretical studies, but these are nevertheless of interest when considering an emerging technology because the ideas described are normally more advanced than those which have been studied experimentally, and a review may help to identify other opportunities.

A comprehensive theoretical study has been undertaken of directly-steering wheels (9) (i.e. the third mechanism mentioned in Section 3), and this identified some of the key issues for stability with such a system. These ideas have been extended in (18) and (19), and in particular have studied the use of differentially controlled traction/braking forces instead of actuators to provide steering action.

For conventional bogies, rather than using actuators directly to provide stability for the wheelsets, another approach is to use active relaxation of the primary yaw suspension (20). The method is very different from active steering schemes which attempt to align the wheelsets radially in curves on the basis of the rotation of the bogies or bodies, because it works only on measurements local to the wheelset. A simple control strategy has been shown to give excellent improvements in curving performance, essentially giving what can be considered to be "perfect steering", and the concept has safety and reliability advantages because dynamic stability is still provided by passive mechanical means, with control action only being applied at low frequencies.

The issue of safety engineering has not been addressed very much, although there was a study some years ago which proposed and analysed a scheme consisting of a diverse 2-out-of-3 redundant controller system (21), secured by voting processors and fail-safe hardware and software. Schemes such as this would need to be incorporated within any practical implementation of an active guidance system, and this work has been referenced to give emphasis to the issue of safety.

## 5. CHALLENGES FOR RESEARCH AND DEVELOPMENT

It will be clear from the previous two sections that many of the important issues relating to active primary suspensions remain to be decided. (This is in contrast to active secondary suspensions, for which most of the fundamental approaches which can be implemented practically have been determined.) There is therefore a clear requirement to undertake more research, and of course the ensuing engineering development which must follow.

A range of mechanical arrangements and associated control options have been suggested, and theoretical studies are necessary to decide what to measure and what control law to apply in order to achieve robust stability for the full range of operational variations in the major wheel/rail interaction parameters which are likely to be encountered, such as effective conicity, creep and adhesion. At the same time these control strategies must respond to curves in a way which minimises flange contact and tread wear. Of course the relative requirements of operational speed and minimum curve radius will depend upon the application, meaning that different wheel mechanisms and control laws may be appropriate for high speed passenger vehicles compared with, for example, light rail vehicles.

A most important issue relates to reliability and safety: for an active secondary suspension it may be permissible to disable the active system after a fault as long as safety is not at stake, but this is most unlikely to be acceptable for an active primary suspension. Consequently the system will need to have some degree of fault tolerance designed-in. Probably the most difficult will be in relation to providing some form of redundancy in the actuation, and the requirements for fault tolerance may have a major influence upon the decision regarding the type of wheel mechanism, perhaps more so than the pure control issues.

It's important to recognise that this issue of fault tolerance in a safety-critical situation has been faced and overcome by the aerospace industry during the successful development and implementation of fly-by-wire aircraft, and without doubt there are significant lessons to be learned from this industry. However there are two major differences. Firstly, the flight control system must continue to operate because it is not possible to provide a mechanical back-up, in an aircraft, whereas for a railway vehicle it is possible to conceive an arrangement in which a mechanical back-up, probably with somewhat reduced performance, is available in the event of a system failure. Secondly, the maintenance/reliability regimes are very different: it is accepted that faults found during pre-flight checks can delay an aircraft's departure, but this is most unlikely to be appropriate for a train unless it is a very infrequent event. There are therefore some interesting and challenging problems to be faced in engineering a practical active primary suspension system, especially in view of the cost-performance requirements for modern railway rolling stock.

## 6. CONCLUSIONS

The paper has provided an overview of the concept of "steer-by-wire" technology for railway vehicles, in which the basic guidance function is provided through the use of active control rather than by the fundamental mechanical system which has been characteristic of railways since their earliest days. A range of possibilities have been explored, some requiring only a relatively small step from current railway practice, others much more radical in their approach.

The title of the paper questions whether this can be a railway technology of the future, but no definite answer is offered by the author. There are undoubtedly technological opportunities and practical benefits, but development of cost-effective systems requires both a strength of purpose to realise the benefits, and great ingenuity and skill to overcome the engineering problems.

## ACKNOWLEDGEMENTS

The author would like to acknowledge funding received to tackle some of the research issues described, from both the UK's Engineering and Physical Sciences Research Council (grant GR/L59795) and the European Commission (Brite-Euram project BE97-4387).

## REFERENCES

1. Karnopp D C: Are active suspensions really necessary?, Paper 78-WA/DE-12 at ASME Winter Meeting, San Francisco 1978.
2. Goodall R M and Kortum W: Active controls in ground transportation - a review of the state of the art and future potential, Vehicle System Dynamics Vol 12, pp 225-257, 1983.
3. Goodall R M, Active Railway Suspensions: Implementation Status and Technological Trends, Vehicle System Dynamics, Vol 28, pp 87-117, 1997.
4. Ford R: Tilt no longer a novelty in Europe. Railway Gazette International, pp 249-252, April 1997.
5. Wickens A H: The dynamics of railway vehicles - from Stephenson to Carter. Proc IMechE Part F (to be published).
6. Gilchrist A O: The long road to solution of the railway hunting and curving problems. Proc IMechE Part F (to be published).
7. Dukkipati R V, Narayana Swamy S and Osman M O M: Independently rotating wheel systems for railways - a state of the art review. Vehicle System Dynamics, Vol 21, pp 297-330, 1993.
8. Pederson J: New S-trains for Copenhagen, European Railway Review, pp 29-33, Nov 1995.
9. Wickens, AH.: Dynamics of actively guided vehicles. Vehicle System Dynamics, Vol.20, No.3-4, pp.219-242, 1991.
10. Leo R: Creep-controlled wheelsets for high speed service - theory and test results. Proc 8th Int Wheelset Congress, Vol 1.2, pp 1-16, Madrid 1985.
11. Chollet H, Maupu J L and Bernard M: ERIOM - Essieu à roues indépendantes orienté magnétiquement. Magnetically steerable wheelset with independent wheels - calculation and tests at scale 1/4. Proc 17th ICTAM Congress of Theoretical and Applied Mechanics, Grenoble 1988.
12. Chollet H, Ayasse J B and Pascal J P: Measurement of the transversal creepforce in a wheel-rail contact area. Proc 11th IAVSD Symp, Vehicle System Dynamics, Vol 18 (Supplement), pp 97-107, 1989.
13. Aknin P, Ayasse J B and Devallez A: Active Steering of railway wheelsets, 12th IAVSD Conference, Lyon, August 1991.
14. Hondius H: Microprocessors harnessed to optimise radial steering. Railway Gazette International, pp 315-6, May 1995.
15. Suda, Y, Anderson, RJ, Yamada, K.: Dynamics of unsymmetric suspension trucks with semi-active control. Proceedings of the 13th IAVSD Symposium on the Dynamics of Vehicles on Roads and on Tracks, Sichuan, China, Aug 23-27 1993, Vehicle System Dynamics, Vol.23, No.SUPPL, pp.480-496, 1994.

16. Jaschinski A et al: Parameter identification and experimental investigations of unconventional railway wheelset design on a scaled roller rig. Proc 14th IAVSD Symp, Vehicle System Dynamics, Vol 25 (Supplement), 1996.
17. Anon: Single-axle running gear - dosaged torque. BahnTech 3/97 (Research and Technology report of DB AG), pp 10-11, 1997.
18. Powell, AJ, Wickens, AH.: Active guidance of railway vehicles using traction motor torque control. Vehicle System Dynamics, Vol.25, Suppl, pp.573-584, 1996.
19. Powell A J, Goodall R M, Walker C, Neut D and Noël Y: Comparison of the mechanical steering system used on the MF88 trains of the Paris Metro with an active guidance system. Int Conf Railtech'98, IMechE, Birmingham, Nov 1998.
20. Shen G and Goodall R M, Active yaw relaxation for improved bogie performance. Vehicle System Dynamics, Vol 28, 1997.
21. Gotz G: Safety engineering for an active hunting controller for bogies in wheel/rail systems. Proc IFAC Symp on Control in Transportation Systems, pp325-331, 1983.

C552/019/98

# A new MEng Degree in railway engineering and systems

**D M LILLEY** BSc, PhD, MICE MIStructE, MIMechE
Department of Civil Engineering, University of Newcastle-upon-Tyne, UK

## SYNOPSIS

This paper provides a basis for discussion of the need for a new undergraduate MEng Degree programme to produce graduates who will subsequently be employed in the railway industry. A suggested outline of the syllabus is included, with the clear intention that it will be modified through discussions and contributions from potential employers and academics from institutions of higher education. A likely timetable for the implementation of the Degree is suggested, and readers are invited to make their views known by returning a questionnaire.

The paper will primarily be of interest to all those who recruit people to work in the railway industry and to those involved with the education and training of students.

## 1 INTRODUCTION

The future well-being of UK industry and citizens is likely to depend on the safe, rapid and economic transfer of goods and passengers over considerable distances with minimum detrimental effect on the environment. Railway developments offer opportunities for improved passenger and freight services, and, with appropriate high-quality infrastructure, could be the most economical and preferred choice for many movements within the UK. Progress will undoubtedly depend on the technical and personal skills of staff at all levels within the railway industry, and the early development of these skills, linked with appropriate depth of knowledge and understanding, is a prime aim of a proposed MEng Degree in Railway Engineering and Systems.

Experience in higher education amongst engineers and people working in the railway industry usually comprises "standard" university degrees formulated to offer studies in a wide range of subjects within a major discipline, eg. mechanical engineering, civil engineering, etc.. Railway-related studies, where available, are sometimes limited to a stand-alone optional module representing less than one tenth of the total study time of that intended within a single academic year. An example of this is the Railway Management, Economics and Planning Module available to postgraduates and final year MEng Civil Engineering students at the University of Newcastle.

A commendable development in railway-related education began in September 1995 when the Universities of Birmingham, Loughborough and Sheffield jointly launched a postgraduate MSc(Eng) Degree in Rail Systems Engineering to provide an integrated overview of railway engineering and management. The MSc(Eng) Degree has its focus within the Advanced Railway Research Centre at the University of Sheffield, and is supported by the IMechE and IEE, and by other organisations active in the rail industry. Full-time or part-time programmes of study are possible, with formal teaching provided between October and March. On successful completion of the taught elements, students undertake a research or development project in one of the co-operating university departments or in industry. Modules are "self-contained", individually assessed, and can contribute to schemes of continuous professional development prescribed by UK engineering institutions.

It is the view of many people that ever-growing pressure in the UK to provide a safe, reliable, fast, comfortable and economically viable method of transport between city-centres cannot be sustained by road traffic; future opportunities for railways in the near- and medium-term future will be substantial and successful implementation will prove vital to the health and wealth of the UK population.

Integration of different aspects of railway operations will be very important in achieving future success, and this will require people working together with knowledge and understanding of how problems inter-relate to other parts of the system and the implications of any potential methods of solution. Management strategies driven primarily by the need to make short-term profit rather than a more long-term view of the operating and maintenance needs of an entire railway system have the potential for a lack of coherence, and may well restrict future progress.

Depth of understanding and knowledge of railway engineering and systems are not currently taught at undergraduate level anywhere in the UK in a way which promotes a coherent overview with later opportunities for specialising in chosen areas. This may be an underlying reason why railways have not progressed and developed as those in neighbouring European countries or Japan. It seems timely to consider whether or not there is a need for a new undergraduate degree, a Master of Engineering (MEng) Degree with Honours in Railway Engineering and Systems.

## 2 COURSE AIMS

A principal aim of an MEng Degree programme is to provide educational study which qualifies graduates for registration as Chartered Engineers. It should provide depth and breadth of coverage to meet the needs of industry in technical, management and business subjects and personal skills. Students should develop a positive attitude towards lifelong learning, probably leading to a wide range of subsequent study. Guidance on the overall content of MEng Degrees has recently been published (1) by the Engineering Council, and this guidance has been broadly followed in developing the course outline described below.

The four-year MEng Degree in Railway Engineering and Systems will aim to produce graduates who are articulate, inquisitive, imaginative, numerate, literate, practical and versatile. Students will develop an understanding and appreciation of railway industry and operations, and the contributions made to improving quality of life and creating wealth. Graduates will be prepared for careers in which they will be able to take leading roles in innovation and change, and able to control projects involving advanced technology with responsibility for management of risk and large budgets.

The development of personal transferable skills such as team-work, professional judgement, and time management will take place at appropriate stages throughout the course. Use of information and communications technology will be taught formally in the early stages of the programme, and developed later through applications and project work.

## 3 LIAISON WITH INDUSTRY

Industrial involvement will be very important in both the design and delivery of the course, and may be further linked through periods of work experience for students and industry-related project work. Practising engineers will be actively encouraged to be involved by delivery of appropriate lectures, internal talks, involvement in design projects, acting as industrial tutors and by allowing students to make visits to appropriate sites and locations. Students will be encouraged to develop links with IMechE, ICE, IStructE, and IEE and to attend appropriate Institution meetings.

It is hoped to establish a "Steering Group" at an early stage with interested parties willing and able to provide detailed guidance on course content and relevance to industrial needs. This group will meet at regular intervals to review the course and to consider evolutionary changes to the programme, when appropriate.

Students will be encouraged to enhance their academic knowledge by gaining appropriate work experience or training during a period of industrial placement and/or during summer vacations. These periods would not normally be less than eight weeks and involve practical design, construction or operation within railway-based industry. Students will be required to produce reports on their experiences during periods of industrial experience. In addition, it is hoped that funding will allow group visits to various railway systems in the UK and Europe to observe different applications and operations of rail transport systems.

## 4 OVERVIEW OF COURSE CONTENT

The content of the course will be constructed around core subjects appropriate to the railway industry, and will embrace theory, analysis and design which integrates infrastructure, rolling stock, power, control and communications, management and economics.

At least half of the course will consist of engineering science and technology subjects involving calculation, experiment, observation and deduction. At least a further 25 per cent of the course will be devoted to subjects directly relevant to engineering but which broaden the students' appreciation of work in the railway industry and the interaction of the engineer with business, society and the natural environment.

Mathematics will be taught within the first two years as a separate discipline, as a means of developing engineering theory, as a method for communicating results, concepts and ideas, and as a tool for solving complex problems.

Experimental and practical work in some subjects (eg. Surveying) will extend beyond laboratory testing and experiments and may be taught partly within residential field courses, if appropriate. Some of the teaching in the later stages of the course will cover current research topics.

A range of non-engineering subjects will be provided for students to follow in varying degrees of depth. Management activities are an integral part of the work of professional engineers, and all students should have the opportunity of learning and understanding the main elements of project control as well as the range of funding options that are available. The teaching of Health and Safety and the need to operate safe systems is essential and will be taught in the first weeks and reinforced at regular intervals throughout the course. This will include legal requirements as well as risk analysis.

Throughout the course students will, through practice and formative assessment, acquire communication skills, including the effective use of Information Technology and the ability to communicate with non-specialists. Graduates will be able to produce appropriate written English that is concise and easily understood.

Students must appreciate that environmental issues are very important to the future success of railways and that an understanding of fundamental material is vital. Railway operations pose environmental problems as well as offering solutions to other problems. Graduates must be able to make decisions based on knowledge of common difficulties and be able to take them into account in railway operations and any related design and construction work. The course will include elements that will allow the development of concepts of environmental sustainability, acceptability and assessment.

In the final year each student will undertake a major investigative project which will address a problem of a non-routine nature. At least one group project will be introduced to develop team-working skills. Projects will normally be linked to a real problem, and will benefit from liaison with industry.

## 5 STUDENT RECRUITMENT AND PROGRESSION

Normal entry to the MEng Degree would be for students with demonstrably high academic achievements and motivation, typically with good A-level passes (Grade B or above) in Mathematics and two other subjects, at least one of which would be science-related (e.g. Physics).

The initial aim would be to recruit about 30 students each year to the Degree; classes of this size offer the advantage of being large enough to cater for a variety of student backgrounds and experiences but sufficiently small that individual students are able to express themselves and develop effective relationships within their peer group and with academic staff and industrial tutors. A prime aim throughout the course will be to provide a first-class learning experience for each student.

It is expected that students would follow rules of progression commonly found within most institutions of higher education. These are likely to require good overall achievement (II.2 Honours level or better) in studies each year before moving to more advanced work in subsequent years. In line with good academic practice, students who cannot complete MEng studies for academic or personal reasons may graduate with BEng (Honours) or BEng Pass Degrees, if appropriate. At this stage it is felt that no direct recruitment would be made to these BEng Degrees, which would be regarded as "safety-nets" having reduced work-loads and academic requirements from those of the MEng.

## 6 ACADEMIC RESOURCES

Academic staff with current railway-related interests are mostly spread throughout several university departments and research organisations, with greater concentrations in places such as Sheffield, Derby and London. In the first instance, creating a new group of academic staff with specific responsibility for promoting and developing undergraduate education in railway engineering within an established UK university may be financially unattractive, particularly in view of the uncertainty about student numbers and current financial problems facing most universities. However, new initiatives in areas which are transport-orientated and sensitive to environmental and commercial needs might benefit from additional funding from central agencies, particularly if strong support were to be forthcoming from industry.

If no single concentration of railway-related teaching resources is possible, an alternative might be to teach different years of the course (or advanced level options) in universities with appropriate staff and facilities. Almost all engineering degree programmes within the UK are now taught in semesters using modular systems; this could facilitate collaboration between academic Departments in different universities. Thus it might be that Years 1 and 2 are taught in, say, Newcastle, and years 3 and 4 in, say, Sheffield (taking advantage of the MSc(Eng) modules already offered by the Advanced Railway Research Centre). Whilst this may appear to be cost-effective, students may find it unattractive because they may feel a lack of "belonging" caused by an unsettling move between universities in mid-studies. Some market research into the likely effects of students moving between universities may prove useful.

It is likely that one university would have to provide a focus for the MEng Degree in Railway Engineering and Systems so that appropriate mechanisms were in place for assessing student progress at the end of each year, ensuring the quality of education, administration, and ultimately to award successful graduates with their Degrees.

## 7 OPPORTUNITIES FOR COMMERCIAL ORGANISATIONS

A prime aim of the new Degree would be to produce graduates who understand the fundamental principles and problems of railways and who are in demand by organisations with interests in the railway industry. This means that the course structure and content would be significantly influenced by the needs of industry and these would need to be reviewed on a regular basis, possibly annually.

It is to be hoped that such organisations would wish to develop an early relationship with students on the Degree programme by one or more of the following:

- temporary paid employment during vacations or periods of industrial placement;
- industrially-related project work in years 3 and 4 in areas relevant to the organisation;
- visits to sites or locations of interest to students where work is in progress;
- becoming members of a Steering Group and/or an Advisory Committee with input to the content, format and structure of the Degree;
- sponsorship of one or more students during their studies through bursaries or scholarships;
- donation of prizes for outstanding student achievement.

Major organisations wishing to make a deeper commitment might consider funding of a "named" academic post, with the intention of gaining further benefit through developing research expertise of immediate interest to the sponsoring organisation. Similar arrangements have been made in other engineering disciplines in some universities. Students would also benefit from the rapid dissemination of results of research funded in this way.

## 8 STUDENT PROSPECTIVE

It is hoped that students would be attracted to the new MEng Degree for a variety of reasons, but principally because it would offer an opportunity to study inter-related subjects of direct relevance to the on-going problem of balancing transport and environmental needs. Commercial advertising means that potential students will be aware that modern high speed trains are becoming more and more comfortable and attractive as a means of transport; they will also know of trains operating without drivers. The potential exists for students to recognise "glamour" in what is becoming a high-tech industry, both in high-speed main-line movements and light rail developments. The links with industry and high probability of employment on successful graduation will also be major attractions. Students might be attracted by the benefits from foreign travel, through visits to observe the operation of European railway systems, and possibly through project work in years 3 and 4 which would complement and develop material taught earlier.

The relevance of the Degree content to potential students would become increasingly evident as cities become more and more congested by road vehicles, with consequent harm to the environment. Some students might find inviting the challenge of developing rail-based solutions for movements to, from, and within city centres.

## 9 ACCREDITATION

The approval of the Institution of Mechanical Engineers and possibly one or more of the other engineering institutions would be sought through professional accreditation of the MEng Degree. Graduates who gain employment in the railway industry should expect to become Chartered Engineers after an appropriate time spent in work-related training and experience, and the award of the MEng Degree should be an exemption from the Engineering Council Part 2 examinations.

## 10 LIKELY TIME-TABLE FOR MENG DEVELOPMENT

The development of a new Degree programme within a UK university is a lengthy process as a result of various tests and checks made to examine its proposed aims, objectives, content, philosophy and outcomes. The process of formation and implementation of the proposed Degree may become further complicated because of the desire to involve industrial partners, the possibility of contributions from different universities and different disciplines of engineering and business management. The likely demand from industry for graduates has been a very important factor in determining the success of any proposed Degree, and this is certainly the case for the MEng in Railway Engineering and Systems.

With thoughts that the construction of 1000 km of track on suitably-prepared ground has to begin with the laying of the first sleeper, a realistic outline time-table for the initial development of the MEng Degree is suggested in Table 1 overleaf.

**Table 1: Outline timetable for development of MEng Degree**

| | |
|---|---|
| November 1998 | Railtech '98 conference. Analysis of questionnaire responses. |
| December 1998 – January 1999 | 1-day meeting of Steering Committee (all interested parties to be invited). |
| January - April 1999 | Other meetings of Steering Committee (as necessary). |
| April 1999 | Completion of planning of syllabus and resources. |
| May 1999 | University Regulations drafted and sent for approval. Information completed for UCAS Handbook, university prospectus, advertising. |
| October 1999 – March 2000 | Potential students apply through UCAS. University selection interviews and processes. |
| August 2000 | Selection processes completed. |
| September 2000 | First students enrolled on MEng Degree. |
| April 2002 – September 2002 | First students undertake industrial placements? |
| June 2004 | First students graduate. |

## 11 SURVEY OF VIEWS

This paper is intended as a tentative first step which may lead to the establishment of a new MEng Degree. It has been written from the viewpoint of an academic with interests in a wide range of engineering matters, some of which are related to railways. There will be many others with valuable views and important contributions to make to the discussion and hopefully, the formation of the structure and content of the MEng Degree programme.

A brief questionnaire (see Appendix II) has been prepared to solicit views from members of the railway industry, academics with railway interests, and any other interested parties. (Loose copies will be available for completion by conference delegates.) The questionnaire includes an invitation to become involved as a member of a Steering Group and/or Advisory Committee. It is hoped that as many responses as possible will be received so that the level of support for this proposed innovation can be accurately gauged. In addition, a national mail-base (mengrail@newcastle.ac.uk) is being established to encourage discussion and rapid transfer of ideas amongst academics and interested parties in the railway industry.

## 12 CLOSING COMMENTS

If the author's belief that the future success of railway-related industry relies upon people with a depth and range of relevant knowledge and experience, then the challenge of developing the MEng Degree will be worthwhile and rewarding. The results from the questionnaire will be of major importance in determining the outcome of this proposal, which is itself believed to be innovative and timely.

## 13 REFERENCE

SARTOR '97, 1997, Standards and Routes to Registration for the Engineering Profession (3rd Edition), The Engineering Council, 10 Maltravers Street, London, WC2R 3ER.

## APPENDIX I SUGGESTED PROGRAMME CONTENT
## MENG DEGREE IN RAILWAY ENGINEERING AND SYSTEMS

### Year 1

Health and Safety: The need for safety, risk assessment, accident statistics, safety principles and processes, the role of the Railway Inspectorate, Health and Safety at Work Act.

History and Development of Railways: The development from the early beginnings of rail-based transport systems, important developments in rail transport, relative costs of rail transport, railway disasters and lessons learned. Recent developments and current plans.

Theory and Design of Structures: Behaviour of simple frameworks and beam structures, methods of analysis, effects of moving loads, design criteria, simple design examples and exercises. Stress analysis and calculation of deflection.

Mechanics: Fundamentals of mechanics of solid bodies, dynamics, motive power and braking systems. Friction, gear systems, power loss.

Materials: Behaviour of materials used in railways: steel, concrete, timber, masonry, copper, alloys, glass reinforced polyesters (GRP), ceramics. Crack formation, fatigue, material defects. Methods of detection and evaluation.

Applied Electricity: principles of electricity, electric motors, simple circuit design, power transmission, fail-safe signalling techniques.

Engineering Mathematics I Calculus: Functions, differentiation and integration. Exponential, logarithmic and hyperbolic functions. Complex numbers, Cartesian and polar forms. Trigonometric functions and inverse functions. Partial differentiation. Techniques of integration and numerical integration. Applications. Introduction to Differential Equations, Matrices and Vectors: Basic types of first and second order ordinary differential equations. Analytical methods of solution. linked systems. Matrix and vector algebra. Solutions methods for systems of linear equations. Eigenvalue problems.

Railway Infrastructure: railway track formation, track construction, bridges, tunnels, points and changeovers, super-elevation of track, curves and transition curves, "hands-on" experience of rail track construction, inspection and maintenance requirements, planned maintenance and methods of urgent repair.

Surveying for Engineers: Simple site surveying and setting out techniques, overview of surveying skills including levelling, angle measurement, distance measurement, control surveys, taping and tacheometry, setting out, curve design, area and volume, photogrammetry. Fieldwork, planning and communication.

Information and Communications Technology: Personal computer hardware; Windows; word-processing, spread-sheets, basic programming languages, real-time control of simple systems for operation and monitoring of equipment. AUTOCAD - Drawing and Editing; Plotting. Visual Basic - Visual Basic Concepts; Designing Programs; Program Flow; Testing and Debugging; Functions; Arrays.

Optional Subject: Chosen from a wide range of subjects offered with the University but might include modern foreign European language, or a subject not directly related to railway engineering.

### Year 2

Infrastructure Design: Layout and conceptual design of stations, depots, maintenance sheds, concourses, franchising.

Applied Soil Mechanics: Role of geotechnical engineering and importance of soil mechanics; formation of engineering soils; soil classification; total and effective stress; pore pressure;

geostatic stresses; stress-strain characteristics; shear strength; seepage; consolidation; compaction; laboratory testing. Stress distribution; limit equilibrium; upper and lower bound theorems; active and passive pressures; retaining structures; gravity, cantilever and embedded walls; stability analysis; types of slope failure, infinite slope failure, circular slope, remedial action, stress distribution; bearing capacity and settlement of shallow foundations. Tunnelling.

Rail Vehicles: Rolling stock: different types, rail gauges, power requirements, braking systems, suspension systems, tilting trains, passenger requirements, freight requirements, design and manufacture of rolling stock.

Hydraulics and Aerodynamics: hydrostatic effects, laminar, turbulent and transitional flows, Bernoulli's equation, drag, aerofoils, lift, aerodynamic braking effects, scour produced by flowing water (safety of bridge structures), sediment processes, cavitation, pumps and drainage requirements. Aerodynamics of trains above and below ground, aerofoil lift and braking, vortex formation. Wind effects. Air pressure variations in underground rail tunnels, implications for fire safety.

Modern Signalling and Communications: Train identification and location, radio links, implications for high speed trains, driverless trains, passenger information systems. Remote monitoring within trains, track, tunnels, etc., etc..

Engineering Mathematics II: Statistics and the Environment: Probability, sampling, descriptive statistics, binomial/Poisson/uniform/ normal/exponential/EVG distributions, linear combination of variables, Central Limit Theorem, t-distribution, confidence interval for mean, comparing means, approximate confidence interval for standard deviation, comparing standard deviations, approximate confidence interval for proportion, comparing proportions, regression on one explanatory variable, multiple regression. Numerical Methods: Ordinary differential equations, linear algebraic systems, numerical integration, introduction to other techniques. Differential Equations and Series: Graphical illustration of solutions of first order differential equations using isoclines. Simple numerical methods of initial value problems. Introduction to Laplace Transforms and their use in the solution of differential equations. Taylor's series for functions of several variables, and stationary values of such functions. Fourier series representation of a function.

Techniques of Structural Assessment: Bridge Inspection Guide. Introduction to Non-Destructive Examination (NDE). Probability applications to NDE. Ultrasonic techniques. Radiography. Use of penetrants.

Optional subject: As for Year 1 above, but a different subject will be chosen in Year 2.

**Years 3 and 4**

Environmental Impact: General effects of railway development. Methods of measuring environmental impact, train-generated sound and vibration, legal requirements, the process of Public Inquiries (with case histories). Passenger needs. The process of Environmental Impact Assessment, EC directives and the UK system, EIA methods, assessment of impacts, prediction and assessment techniques, monitoring and impact auditing, risk assessment, English Nature & EIA, case studies.

Repair and Refurbishment of Railway Infrastructure: Refurbishment with respect to railway structures and property. Assessment of structures for refurbishment; original document recovery and interpretation; design of temporary support to facades; foundation underpinning; case studies.

Scheduling: Practical considerations, train, track and staff availability, passenger demand forecasting, cost benefit analysis. Methods of timetable/schedule formulation. Turn-around times. Power and fuel considerations. Passenger interfacing. Ticketting.

Railway Economics and Marketing: Marketing techniques, micro- and macro-economics of railway systems: inter-city routes, light rail systems, freight handling, mixed road-rail systems.
Methods of Modern Management: Material resource management: cost-efficiency and benefits, just-in-time approach, crisis management, priority allocation. Man management: importance of motivation, recognition and reward of personnel. External accreditation, e.g. Charter Mark, Investors in People, etc.
The Railway Engineer in Society: Professional ethics; questioning; model of a value-system; welfare. The public and the professionals. Relations between Engineers, Planners, and Architects. Interactions within design and project teams. Case histories. Assessment of the effects of projects on the environment. Insurance and litigation. Engineering contracts, responsibility and liability.
Economics for Engineers: The market environment for decision making, the firm in short-run and long-run, discounting techniques, comparative analysis of investment criteria, cash flow and inflation, taxation, opportunity cost, replacement repair, capital budgets, capital markets, cost of capital and risk, international aspects of project appraisal, basic consumer theory, object and basis of cost benefit analysis, measuring welfare changes and shadow pricing, direct and indirect effects and consumer surplus, non-marketed goods, including externalities and public goods, the discount rate for public projects, distributional weights.
Strategic Management and Organisational Theory: Mission statements, goals, policies, tactics, effectiveness -v- efficiency. The determinants of strategic position, the environment, resources, objectives and expectations, life-cycle, value-chain, generic strategies, strategic choice, evaluation and selection of alternatives, technology management, problems of implementation. Strategic case study. Scientific management, theory X and theory Y, Eltom Mayo and the Hawthorne investigations, socio-technical systems. William Ouchi and theory Z. Charles Lindblom and the "Science of Muddling Through", Henry Mintzberg's "The Manager's Job: Folklore and Fact", Mintzberg's "Structure in Fives", a case study in organisational decision making.
Project: Each student will undertake a substantial project in collaboration with an organisation working in the railway industry. Ideally, the project will address an existing problem or difficulty or examine potential issues arising from future activity relevant to the collaborating organisation. Each student will be supervised by an academic and/or a member of the industrial organisation, as appropriate.

Other subjects will be incorporated following discussions with academics and taking into account the views of the Steering Group or Advisory Committee.

C552/004/98

# UTILITY– utilization of transport infrastructure and land through information technology

**F C PERROTT** MIMechE
Any System Limited, Cirencester, UK
**A C RENFREW**
Department of Electrical Engineering and Electronics, UMIST, Manchester, UK

## SYNOPSIS

The drive to transfer urban traffic from road to rail may demand costly and intrusive infrastructure. This paper reviews the use of emerging technology to increase existing corridor capacity through the operation of contacting or closely-spaced vehicle convoys, a concept in which automotive manufacturers are also showing interest. The authors view the '1-dimensional' nature of rail transport as a realisable environment and proving ground for technology transfer of headway control to non-guided systems. The paper considers the features, problems and technology involved in status monitoring of rail vehicles operated in a 'no block' convoy and looks forward to future applications.

## 1. INTRODUCTION

Some urban road and rail infrastructures are already unable to accept all the traffic which is on offer at peak periods. Although it is perhaps inevitable that new infrastructure construction is necessary, such new construction is more readily justified if it can be shown to catalyse improved service quality and efficiency in the use of land. The authors' aim is to make up 50% of track space simultaneously available for occupation by revenue-earning vehicles at the required speed. For both rail and motorway, 5% is a closer estimate of current performance and represents a significant waste of land and investment.

For rail, this figure is based on an underground route of capacity 30000 passengers per hour, travelling at 56 km/h and with a density of 14 passengers per linear metre of train; approximately 4% of track length is occupied by revenue earning passengers at one time. For a three lane motorway with a nominal capacity of 10000 passengers per hour (at an average of 1.5 per car) and a car 'space' of $12m^2$, approximately 5% of the road area is utilised by passengers.

To make the best use of urban railways to relieve road congestion, some realignment of tracks, updating of signalling systems and relocation of stations may increase the throughput of traffic. However, to increase passenger appeal, i.e. woo the passenger from his/her car, an alternative strategy which brings journey flexibility closer to the 'door to door, no interchange' ideal scenario may be demanded. The 'Utility' operating strategy is intended to answer this requirement by combining self-routing vehicles and closely spaced or contact 'trains' of vehicles having no pre-determined maximum 'block length.' Although it dates back over thirty years, the convoy principle may now be approaching commercial feasibility, thanks to developments in communication technology and sensor-associated software. The control of train length in respect of loading/unloading constraints is considered to be a matter for control by network software rather than signalling hardware.

## 2. TRACK HARDWARE

Although the starting point for this research is close to established rail technology, the emphasis upon improved passenger access, reduced track costs and minimised intrusion infers that the technology, once proven, could be equally applied to 'road' systems, i.e. those without fixed hardware guidance or any hybrid design which might arise in the future. A research programme would expect to be linked to existing research and evolving technology for road and air systems.

A range of types of rail hardware may be used with a 'Utility' operating strategy. For conventional rail track, self-routing vehicles may diverge or converge using an existing point design (1) which has no moving parts, thus avoiding the need for clearance times and the switching associated with certain Personal Rapid Transit schemes. Figure 1(b) shows a model of this track selection junction; the horizontal 'selector wheel' may be engaged with either of the two higher level horizontal rails, thus constraining motion to left or right. Figure 1(a) is a full size installation at British Steel, Scunthorpe and allows both vehicle route selection and low-speed conventional switching, by the use of two hinged rails which bear on the back surfaces of weight carrying wheels. Although further development of the design might allow the incorporation of conventional switching, the authors do not consider that 'Utility' and conventional trains could readily co-exist, in view of the need for separation and duplicated control systems. Elevated systems, using lightweight structure and track designs, may be constructed alongside or above railways or roads; monorail or gyro-stabilised systems (which allow narrow tracks) are also possible. In all cases, the enhancement of 'customer friendliness' is due to the self-routing nature of the vehicles (1) and that of improved capacity to the operating strategy.

As an example of the capacity available from a given area, a double track railway of 4.5m width, if using contact convoys of vehicles, would have a greater capacity than a motorway of 33m width.

## 3. 'NO BLOCK' CONTROL AND STATUS MONITORING

Instead of locating each coupled train within a fixed or moving block, status monitoring is used to individually programme each vehicle. It then monitors the compliance of each separately and virtually continuously for speed, position and acceleration/deceleration. An inherently safe closed-loop control system eliminates much of the need for longitudinal clearances between individual vehicles, though the requirements for resolution and parameter identification are eased by the provision of 'bridging dampers' which allow small transient variations in longitudinal position to occur. A range of sensing mechanisms is now available, from incremental markers on the track to satellite positioning and optical image processing techniques. In view of the safety critical nature of public transport and the need for high reliability of service, it is important that redundancy and condition monitoring be incorporated.

At first sight, it would appear that there would be no need for any interstice between either consecutive vehicles or convoys. However, as instantaneous acceleration or deceleration of fresh vehicles joining convoys or vehicles diverging cannot take place, some clearance is necessary. Such gaps can subsequently be closed up and a nominal 50% allowance is provided in system design. For example, consider four-seater vehicles (1), travelling at 100km/h, each 3.5m in length, carrying an average of three passengers, and having 50% track capacity allowance, i.e. 3.5m between each pair of vehicles. The peak throughput of passengers would be 43000 per hour but would be less in practice, to allow operational flexibility in a multi-route system. For a relatively unobtrusive system, this figure of track utilisation exceeds that currently realisable with moving blocks. It would not be necessary for vehicles to depart singly from an off-line station; if an appropriate network control system were employed, departures in contact convoy form may optimise network utilisation. Track 'utilisation' is here considered in terms of usefully moving traffic only, as opposed to 'occupancy,' which includes stationary vehicles. The design of stations is also considered important from the viewpoint of land utilisation: at relatively high speeds (100km/h, say), a high capacity station may occupy a substantially larger area than would be the case at lower speed; hence the authors have assumed that stations will not normally occur on through running lines and will be approached or left at reduced speed.

Individual vehicles may join or leave convoys as required (1). The flexibility afforded allows convenience and efficiency in operation, particularly when synchronous speed changes can be applied to an entire convoy, combined with reduction of the delay encountered in a fixed-schedule environment when a peak of traffic occurs. The authors have assumed the use of small, compact vehicles which would reduce visual intrusion, track costs, axle loading and infrastructure wear.

Within a convoy, the bridging dampers mentioned above stabilise transient movements during normal operation. Unless two convoys are joining or separating, there will normally be a substantial clearance between them, calculated according to speed and braking profiles. When manoeuvring to form or split a convoy, the control demands placed on adjacent vehicles may vary. Figure 2 illustrates the procedure of forming a concatenated convoy after a convergence.

Through a telecommunications link between an on-board computer and trackside/centralised processors, every vehicle has access to information relating to the present and intended future movements of those in front of it. There is no necessity to form a convoy unless capacity considerations make this preferable. For example, the arrival of a large airliner, or the end of an entertainment or sport event, may render the use of shuttle convoys essential in maximising departures.

A Master Control Unit will hold command programmes for all possible journeys. Before a particular journey commences, and/or at intervals thereafter, the appropriate command programme will be copied to the computer on board the vehicle (Vehicle Control Unit), rather in the manner of an executable electronic token, but without manual initiation. Thus two copies are held, one in each computer. On a real time basis, using synchronised clocks, demanded 'command status' information (speed, position, acceleration) is provided for closed-loop control, using 'actual status' data for feedback. Although the vehicle will continue to execute its programmed demands, its status will be transmitted at intervals and monitored via telecommunication links, incorporating redundancy, to the Master Control Unit, which will meet standards for safety critical equipment. Polling of each vehicle every 5 seconds can be accomplished with up to 500 vehicles per Master Control Unit (2). In the relevant communication protocol, the Vehicle Control Unit may request a change of status, based on the information at its disposal, as obtained from stored data and sensors. As the Master Control Unit will have access to larger pool of information, permission to execute the change of status may be granted or denied.

In the event of a reported but undemanded change of status which the Vehicle Control Unit is unable to correct, for example due to partial loss of data or adhesion problems, the Master Control Unit may amend the demand signals to all Vehicle Control Units within the affected area. If the deviation exceeds a permitted band, due, for example, to total power or communication loss or damage, a 'scream' channel, separate from the communication uplink and downlink (2), would be provided for controlled emergency shutdown of all vehicles in the sector concerned. A modular control system with fault tolerance and vehicle movement strategy has been developed (2) and attention has since turned to consideration of the Network Control Unit.

Network control, including fleet optimisation, routing, diversion and ordered restarting is the task of a non-safety critical Network Control Unit which also provides user interfaces for operating staff and customers. Recent work has been directed towards this aspect. It is envisaged that 'virtual convoys' would initially operate to a pre-set schedule, whether any revenue-earning traffic is carried or not. When traffic is on offer, or dispersion of the fleet is required, real vehicles would be allocated to those convoys. This is not, however, part of the safety-critical element, as it would be in a moving-block signalling scheme. The number of vehicles in each 'virtual convoy' will vary with demand and the schedule will similarly be varied as the Network Control Unit "learns" to predict demand and fleet distribution requirements.

## 4. SPEED CHANGES

The authors envisage that the Network Control Unit will not directly regulate vehicle speed, though it may specify arrival times. The Master Control Unit will specify speed commands in steps only, a maximum speed being set for each section of line but capable of reduction in steps for traffic regulation purposes. Typically, the maximum rated speeds for an urban transit would be 100km/h, with 50km/h in congested areas and, say, 25 km/h for distribution lines. Finer steps would be applied for curves and junctions and for the closing up of vehicles within a convoy. It will be the task of the Vehicle Control Unit to meet the speed demands by attaining a speed/position/time profile which is substantially within the traction and braking capability of the vehicle, allowing for load, gradient, resistance and adhesion factors. Each vehicle will modify its profile according to its knowledge of the existence of vehicles in near-contact and thus maintain contact position within the limits of the bridging dampers.

## 5. VEHICLES AND POSSIBLE APPLICATIONS

The choice of vehicle for a guided system will depend fundamentally on the system environment, i.e. whether existing 'heavy rail,' semi-buried or elevated infrastructure is most appropriate, and whether visual intrusion is a dominant factor. A typical 'small' vehicle (Figure 3), carrying perhaps 4 passengers and offering rapid response to control demands was used in initial system design studies and has the added prospect of having traction equipment based on available industrial design. A more radical scheme might even be based on the very old (and proven) concept of a gyroscopically stabilised narrow-track vehicle (3), which would allow minimum track cost, simple on-board track selection and the prospect of flywheel energy storage without additional equipment cost. Continuing down the route of possible applications of the control and sensor technologies which will be developed for convoy operation, the prospect of technology transfer to lane-constrained road vehicles would similarly allow greater capacity on, say, a motorway, once a communication channel (using leaky coaxial cable, for example) and software were sufficiently proven to remove the artificial restriction of the 'brick wall' stopping distance (2). As certain modern aircraft are unstable in flight but maintained in safe attitude only by their control loops, the authors consider that control systems have already proved that technology can transcend the 'practical limit' and look forward to sharing research results in future.

## 6. CONCLUSION

'Utility' is a methodology for enhancing both customer choice and the capacity of existing infrastructure, principally through investment in sensor, electronic and information technology. The authors consider that 'Utility' offers further scope for development of innovative intermodal technology and also a proving ground for technology transfer to non-guided systems.

## 7. ACKNOWLEDGEMENTS

The authors would like to thank Dr David Brader of Viewpoint Systems Limited, Manchester, for permission to publish specific information on the proposed control system.

## REFERENCES


1. Perrott, F.C. 'Convoy Rail. Is this something really new?' Newcomen Society, Evolution of Modern Transport, New Series November 1993

2. Brader, D.J. 'Computer Control of Automated Transit,' PhD Thesis, University of Manchester, 1985.

3. Brennan, Louis 'Gyroscopic Monorail,' Gillingham Public Libraries, Local History Series No. 5, part 2.

Figure 1(a) - Full size demonstration point at British Steel, Scunthorpe, with provision for passage of conventional vehicles

Figure 1(b) - Model of point for self-routing systems with track selection mechanism

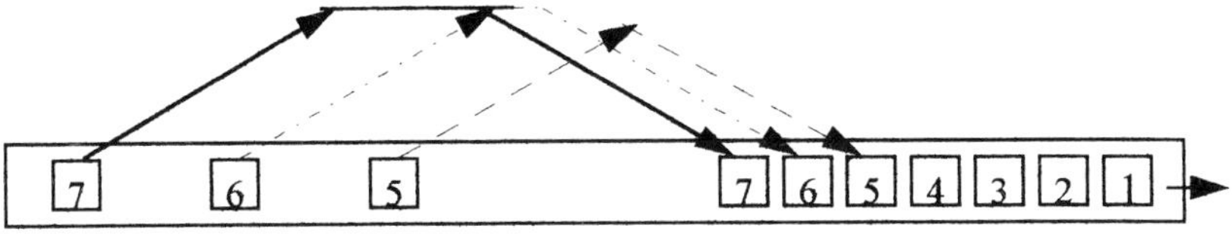


Figure 2 - Closing up of vehicles to form a convoy

Figure 3 - Form of 4-seater vehicle used in initial system design studies

C552/033/98

# The use of simulators and CBT for railway driver training

**J W SCOTT** IEng, MIMechIE
Orchard Consultancy Limited, Derby, UK

**SYNOPSIS**
In this paper, the author examines the use of simulators and computer-based training (CBT) for training railway drivers. The introduction looks briefly at railway driver training simulator history, and typical features found in them. Some of the factors that may determine the use of simulator training are considered. Certain currently available simulation technologies, and their contribution to making simulators effective and efficient training tools for railway applications, are examined. The forms which a simulator training facility may take are also discussed. The appropriate provision of driver training simulators is explored, with reference to specification and procurement issues, and in conclusion the advantages of employing the technology are summarised.

## 1. INTRODUCTION

### 1.1 What is a driver training simulator?

For many years, aircraft pilots, both military and civilian, have undertaken at least part of their training on flight simulators. The high capital value of aircraft and their uneconomical use for non-revenue training flights have necessitated a substitute being found for training purposes. Financial and safety risks demand alternative strategies for training in other similar situations; hence operatives of hazardous plant such as chemical works or nuclear power stations are also being trained on simulators.

Railways have, by comparison, been recent entrants into the field of simulator training for drivers. The increasing complexity of traction and rolling stock, the development of train driving cab design, and the need to put the trains into service quickly and efficiently have all led to the use of simulators as a cost-effective training solution.

Railway driver training simulators are a means of training drivers "off-line" in a risk-free environment. Simulators enable drivers to be exposed to fault, failure and emergency scenarios in complete safety, with the opportunity to review their actions and reactions in exercise replay modes. The simulator records, reports and stores numerical scores for trainees' performance, and allows instructors to use identical exercises for different groups of trainees, thus ensuring a degree of uniformity in training quality and content.

Computer-based training (CBT) uses the power of desktop computers, networked together, to deliver training exercises to trainees, enabling them to work at their own pace and giving them a degree of autonomy in the rotation and choice of exercises. A dedicated workstation allows the instructor to check trainees' progress, offer assistance and advice and maintain records of successful exercise completion by trainees.

### 1.2 History

The earliest recorded use of simulators for railway driver training in Britain that the author has discovered so far was for the former British Railways (BR) West Coast Main Line electrification in the early 1960's [1]. The simulator used a full-size cab mock-up, and a ciné-film of the route was projected onto the windscreen. Few other details are known about this interesting pioneering attempt in this field.

In France, in the late 1980's, two serious railway accidents resulted in the decision by SNCF to invest in driver training simulators, to train drivers especially in fault and failure rectification and emergency procedures. Thirty-six single-cab simulators were built, and delivered to depots over the entire network, allowing training to be undertaken locally. Other railways world-wide were also using simulators at this time, to meet specific driver training needs. They are especially popular for some of the intensively used metros and urban railways in the Far East. Other railway driver training simulators are known to be in use, or to have been used, in America, Australia, and former Eastern Bloc countries.

### 1.3 Recent developments in Britain

In 1989 a pair of driver training simulators were built for the newly introduced BR Regional Railways Class 158 Diesel Multiple Units. These simulators were housed in two road motor coaches, which enabled them to be driven around the country to the various depots from which the Class 158s would work, thus taking the training facility to the drivers. In 1991, a two-cab simulator was delivered to BR's Network SouthEast for training drivers for the Class 465 Electric Multiple Units about to be delivered. This simulator was located in a purpose built training room at the Waterloo Training Centre, London. At about the same time, London Underground took delivery of a driver training simulator for their new Central Line trains. This two-cab simulator was built into a standard 40-foot ISO container, allowing it to be moved, should the need ever arise.

Other driver training simulators in Britain were supplied for European Passenger Services "Eurostar" trains, Channel Tunnel Class 92 Freight Locomotives, Eurotunnel "Le Shuttle" trains, and BR Regional Railways' Class 323 Electric Multiple Units. More recently London Underground has procured simulators and CBT for their new Jubilee Line and Northern Line rolling stock.

### 1.4 Typical features of a driver training simulator

Driver training simulators usually include the following features, although some may be omitted in certain circumstances, depending on the purpose and scope of the simulator.

- An instructor's console from which the progress of the trainee and training exercise can be monitored and controlled;

- A computer network which allows the various parts of the simulator to interact without incurring time penalties;
- Information libraries (databases) containing models of the route, signalling, train systems and performance, and inter-acting systems;
- A realistic driving environment (cab), with all necessary controls active;
- A vision system which displays the view through the cab windows and windscreen, as seen by the driver;
- A sound system which reproduces the sounds made within and by the train;
- A means of recording and storing trainees' performance logs for subsequent analysis.

In addition, some simulators are fitted with systems to impart motion to the entire cab, or more limited representative motion to the driver's seat.

## 2. IS SIMULATOR TRAINING APPROPRIATE?

The acquisition of driver training simulators and CBT suites is not a trivial commitment for most training department budgets. In order to justify the investment, the need must be supported by a robust business case. However, by adopting a systematic approach to identifying the requirements, sound financial foundations can be laid to underpin the procurement of the simulator and supporting training equipment. There are two elements to consider: the process for defining the need, and the circumstances which dictate the need.

### 2.1 The process

It is essential that the training department is clear about the direction and scope of driver training to be undertaken. It will only be possible to see if simulators and CBT represent the best training media when the training needs are clearly defined. Therefore it is necessary to carry out at least two analyses to establish the optimum provision to meet requirements: a training needs analysis, and a training media analysis.

The training needs analysis should identify the needs of the training department and of the trainees. The training department must be clear whether it intends to educate the drivers as well as train them, that is, whether there will be elements of subject teaching in addition to skills training. It is also important to know the aptitude and academic competency profile of the trainee body, in order to tailor the provision of courses appropriately. The training needs analysis will reveal the areas on which emphasis must be placed to meet the aims and objectives of the training department.

The training media analysis will assist in determining the most suitable media for the successful delivery of training. It should take account of the results of the training needs analysis, and consider the full range of options available. Ideally, it should be conducted by an independent body with no commercial interests in the supply of training equipment, and no connection with the training department. In this way, impartiality will be seen to be maintained, without compromising efficiency, quality or economy.

### 2.2 The circumstances

In Britain, training is a legal requirement for employees [2], and on railways in Britain the training of drivers is a mandatory requirement under Railtrack Group Standards [3]. Train Operators also need to demonstrate in their Railway Safety Case that their drivers are competent. There may be other, commercial reasons, quite separate from those above, for providing new training for drivers. An example would be "Total Route Modernisation", usually featuring the provision of new rolling stock, new signalling, new infrastructure and new services. Expanding these subjects, the following sub-groups emerge, any of which, individually or in combination, will require the provision of training:

- Train handling
- Signalling system operation
- Fault and failure procedures
- Safety and emergency situations
- Route or service familiarisation.

There may be other factors, which compound the need to employ driver training methods other than the conventional "Master/Apprentice" approach. For example, traffic density on available routes combined with the number of drivers to be trained may make it difficult to deliver adequate training within a realistic time frame. Reductions in track access charges and training costs may be being sought, or training to drive rolling stock which has not yet been cleared for running on the railway network. These circumstances may direct consideration to the provision of driver training using simulators and CBT, as a cost-effective means to deliver adequate, timely training.

## 3. ADVANTAGEOUS FEATURES OF SIMULATION TECHNOLOGY

Computer and simulation technology have recently developed very quickly. Many of those developments are the result of military, business and leisure-market demands. For civil simulation the spin-off benefits mean superior quality training tools are available at affordable prices, with features which make them increasingly attractive. Rising volume production and emerging industry standards in computing contribute to the ready availability of reliable equipment, with minimal need for bespoke hardware.

The inherent features of computer technology contribute to the competitiveness of simulators as effective training tools. These include processing capacity and speed, the ability to interrogate for numerical data, and the ability for repetition with mathematical precision of an accuracy otherwise unattainable. Thus the reporting, recording and reproduction of trainees' actions and reactions is almost indisputable, and is certainly unbiased by sympathy or prejudice. Furthermore, these reports and records can be generated and stored easily, without recourse to manual inputs other than personal identity. This means that the integrity of the records is maintained, and readily accessible to those with legitimate reason.

There have recently been developments of certain specific aspects of simulation which merit closer consideration: computer generated images (CGI) for vision systems; audio generation and reproduction; and motion systems. These represent three of the most important inputs to the trainee driver; getting them right is essential if the simulator is not to be devalued as a training tool.

### 3.1 CGI vision systems

The European Passenger Services "Eurostar" driver training simulator used an early example of CGI, which was an excellent application of the technology, bearing in mind the length and complexity of the route. However, contemporary railway driver training simulators still tended to use video-recording based imaging systems for the "out of the window" view, due to the prevailing high cost of CGI systems. Military and civil aircraft simulators used CGI extensively, but the images portrayed were often very basic and simple compared to the railway environment. Developments in this technology, and reductions in costs, have led to the use of CGI in more recent railway simulators. Some extremely realistic examples of computer generated railway routes have been demonstrated, and CGI enjoys the following advantages over video-recorded route images:

- Ease of alteration of, for example, new track layouts or new signal positions.
- Weather conditions do not prevent image capture.
- "On-line" controllability of weather conditions, visibility and time of day.

There are, however, disadvantages, of which the biggest is possibly the volume of data to be supplied to produce a convincing picture. Typically, this would consist of visual information about colours and textures, and precise locations of significant elements such as signals, points and platforms. This information would be automatically captured on video recordings. Additionally, it is important that the CGI artists have an appreciation of the railway environment if irritating errors of detail are to be avoided.

### 3.2 Audio cues

The accuracy of the audio cues in a simulator training environments is almost as important as the accuracy of the visual cues. Audio cues in a simulator fall broadly into two categories: sounds and noises. "Sounds" are those audio cues which are essential to enable the trainee driver to follow the correct procedures, for example alarms and warnings, voice prompts and the sounds generated by equipment operating correctly, or malfunctioning. "Noises" are those incidental audio cues which contribute to the realism of the simulator environment, but which do not constitute essential information. Examples of these would be the sound of running over bridges and into stations, or power-operated doors opening and closing.

Some audio cues can only be recorded from real equipment or real-life situations, but some in-cab equipment can be used to generate their own sounds. Care needs to be exercised to ensure that sound reproduction from real equipment and recorded sources is sufficiently well integrated as to maintain the realism of the training simulator environment. Modern digital recording techniques allow the recorded sounds to be manipulated and reproduced with a clarity and accuracy only limited by the quality of the speakers. Time spent ensuring that the audio cues are clearly and accurately reproduced will assist in establishing the credibility of the simulator.

### 3.3 Motion

The issue of motion for railway driver training simulators has been debated at length, but again, cost has been a deciding factor. Motion platforms with six degrees of freedom are expensive, but whilst they may be essential for fighter aircraft training simulators, the value of such complex equipment for railway simulators has been questioned. Basically, there are three options as far as motion in railway driver training simulators is concerned:

- No motion at all;
- A motion seat, imparting a limited degree of movement to the driver's seat only;
- A full motion platform, imparting motion to the entire cab.

Frequently, the only choice considered for motion in railway driver training simulators has been a full motion platform. When that not inconsiderable cost has been examined, the option has generally been rejected, as not having sufficient training value to justify it. However, the lack of motion in a simulator can often be distracting in itself, as the body expects certain motion inputs in response to visual cues. When these inputs do not occur, distressing "motion sickness" can result, but many driver training simulators without any motion system have been used very successfully.

Motion seats represent a low-cost alternative to full cab motion systems. They usually only impart very small movements to the driving seat itself. Nevertheless, problems can occur if the occupant of the driving seat is also in contact with a fixed part of the cab. The relative movement will again be unrealistic, and may send the wrong signals to the person's sensory system.

Motion platforms support the entire cab, and usually the vision projection system also, and move the cab through quite large displacements, in order to impart sensations of acceleration, braking and curving. Until recently, hydraulic drives have generally been employed, with their attendant costs of pumps, pipes and controls. Pumps are usually sited away from the simulator cab, as they tend to generate unwanted noise, and even the drive systems can contribute to the undesirable background noise in simulator rooms. Developments about five years ago saw the emergence of novel electric drives, which, by their nature, tend to be quieter in operation than hydraulic systems, and on the whole, are cheaper. Combined hydraulic and electric drives have been demonstrated, where the dead weight of the cab is supported in equilibrium by hydraulic rams, and the displacements are effected by electric actuators. Personal experience of driving a full-motion simulator suggests that the increased realism is beneficial, but the increased cost may still be prohibitive. Certain manufacturers are claiming low-cost motion systems [4] and this is an area where developments are still taking place; eventually, such systems may become wholly affordable and specified as a matter of course.

## 4. SIMULATOR TRAINING FACILITY OPTIONS

### 4.1 Configuration

The configuration of a driver training simulator will depend very much on the duty which it is called on to perform. As has been noted, the Class 158 simulators were fitted into road motor coaches, to give them a high degree of flexibility in use. They were basic train handling simulators, and performed their function perfectly well, which was to train a geographically widely separated workforce on new traction. The "Eurostar" driver training simulator was, by contrast, something of a flagship installation, representing state of the art technology, and a considerable investment to European Passenger Services. The comparative complexity of the drivers' tasks on these two train was reflected in the two very different approaches adopted.

There is an almost infinite number of simulator training options and configurations possible. Briefly, the established options are:

- Computer Based Training, using desk-top computers under instructors' supervision;
- Part Task Trainers, used for training specific procedures or equipment handling;
- Full Cab Simulators, to train drivers in a realistic "risk-free" environment.

One recent driver training simulator suite in Britain successfully integrated all three elements, although only the Part Task Trainer and the Full Cab Simulators were actually networked together. Clearly, the constraints of cost, time and space, and the purpose for selecting simulator training, will have a direct bearing on the configuration finally selected. The output from the training media analysis will determine much of the detail of the equipment to be selected.

### 4.2 Portability

Again, there will be a number of factors determining the degree of portability of the driver training simulator facility. They will largely be practical considerations including access, staff geographical distribution, and security. Where computing equipment is concentrated in one place, and that equipment is compatible with domestic computers, there will always be a security risk.

Permanent locations have much to commend them from a security point of view, as well as enabling the facility to become "established". Where space is not a problem this is obviously the best option, and where any kind of motion feature is included in the simulator, it is almost mandatory. Integration with other training resources on the same site is straightforward, and frequently desirable.

Transportable simulators are usually nominally permanently sited, but constructed to permit their re-location with minimum disruption. The London Underground Central Line simulator mentioned above, is an example of this type. By being built into a standard ISO container, it was possible to transport it to a new site on a flat-bed articulated lorry.

Mobile simulators, built into road motor coach or lorry bodies, or road vehicle trailers, have been used in numerous military and civil applications. The BR Regional Railways Class 158 simulators referred to above exemplify this type. Where it is not possible to take groups of trainees to one location, then the option of taking the simulator to them becomes the preferred choice.

There are two factors which need to be considered if transportable or mobile configurations are to be adopted. Road vehicles are restricted in size, and this can lead to difficulties in accommodating all the features deemed to be desirable in the simulator. There may also be problems if the trainee group is numerically large, as they may not all be able to observe a training exercise simultaneously. Computing equipment needs to be specially housed if it is to be moved frequently. Those persons responsible for moving the simulator should ensure that it is properly prepared for transit at the appropriate time.

### 4.3 Size

The physical size constraints of road vehicles on transportable or mobile simulators have been alluded to above. The actual size of the facility, regardless of its portability, will be determined by other factors also. These may include the scope of simulation, the number of different types of traction units to be simulated, and the number of cabs required.

The scope of simulation will cover issues such as how much of the train will be physically represented, and which controls will be operational. A half-width cab, for example, with only the primary controls operational, will occupy far less space than a full width cab with external platform monitors, and all controls operational. The number of operational controls will also have a bearing on the space occupied by the associated computer equipment.

Where more than one type of traction unit is to be simulated, it will be necessary to procure at least one simulator cab for each type. Although certain components of the simulator databases and instructors command and control centre may be common to all cabs, there will still be a considerable space requirement for additional cabs, as well as the peripheral equipment accommodation.

It is rarely practical to order a simulator with fewer than two cabs, since this ensures a degree of redundancy as well as representing better usage of simulator time and human resources. In some cases it is essential to have two cabs, to simulate rescue scenarios such as the "push-out" procedure adopted by London Underground. Some physical separation of the cabs is desirable, especially if they are not soundproof, since sounds from one cab will "spill over" to the other.

## 5. ENSURING APPROPRIATE PROVISION

The specification and procurement of driver training simulators are areas in which great care must be exercised if the training tool delivered is to perform the tasks required of it. There are a number of misconceptions surrounding driver training simulators which do not help in establishing them as serious training tools. By considering some of these issues, it may be possible to avoid costly errors, both in terms of monetary and training value.

### 5.1 Involving the end users

It cannot be too strongly emphasised that the end-users should be involved in key stages in the procurement of the simulator. Training department staff, especially those responsible for training development and delivery, should have a significant input to the business case for the procurement of simulators and CBT equipment. They know the training needs, the best delivery media, and the capabilities and limitations of those media. They are also aware of the training programme content requirements and time constraints imposed by the introduction of new training equipment. They should be involved in the development of the specification for the simulator, the selection of a suitable contractor, and in the approval of the design as it emerges. This involvement will assist in avoiding the inclusion of unnecessary features, and the omission of important ones.

### 5.2 Maintaining the focus

Care must be exercised in deciding what to include in the simulator if unnecessary expense is to be avoided. There is a temptation to have the driver training simulator replicate all features and functions of the real train. This is not always necessary or practical, and can lead to over-complication of the simulator. In addition, there will inevitably be some training activities which can only be carried out on the real trains. For example, windscreen wiper failure will render a real train unfit for service. Most driver training simulators have some form of weather condition control, including simulated rain, but few, if any, simulators are fitted with operating windscreen wipers, for numerous practicable reasons. Hence, specifying "Windscreen Wiper Failure" as a fault condition on the simulator would be almost pointless. The simulator ought to be designed primarily to assist training of those events, situations or procedures which may be difficult, dangerous or impossible to reproduce on the operating railway.

### 5.3 Health and safety

Consideration should be given to the psychological and physiological aspects of using simulators to train drivers. On the whole, simulation technology is mature, and most risks are known and can be adequately managed. There are some features of simulators which need special attention if they are not to become stumbling blocks to effective training, and the issue of the integration of motion and visual cues is especially important.

## 6. CONCLUSION

In conclusion, the advantages of simulator training for railway drivers can be summarised as follows:

- It is safe;
- It is almost entirely risk-free;
- It is easy to repeat exercises with a significant degree of inherent accuracy;
- It is cost-effective;
- It is demonstrably consistent;
- It is unaffected by sympathy or prejudice;
- It is nominally instantly available.

However, attention must be given to the following points to ensure that the simulator represents the best solution for the training need:

- A robust business case for employing the technology;
- A clear specification based on training needs and training media analyses;
- The involvement of the end users in the procurement process;
- The exclusion of all unnecessary features and functions.

The future of railway driver training using simulators will depend largely on the recognition of the part which the technology has to offer. As procurement time scales for rolling stock become compressed, and the need to put trains into service more quickly brings pressure to bear on training departments, then ways of accelerating the training processes will need to be sought. Driver training simulators offer significant contributions to the achievement of those ends, and will continue to improve with the ongoing development of computing technology.

---

## REFERENCES


[1] Webb, B. and Duncan, J. "AC Electric Locomotives of British Rail" David & Charles, Newton Abbot, 1979.

[2] The Health and Safety at Work Act 1974, Sections 2 (1), 2 (2) and 7.

[3] Railway Group Standard GO/RT 3251 "Safety Requirements for Train Drivers" Issue 1, Railtrack plc Safety and Standards Directorate, September 1994

[4] Wierda, G.J.,(1997) Fokker Control Systems "Low cost motion systems for ground based vehicle simulators" Proceedings of the 8$^{th}$ International Training and Education Conference, Lausanne, Switzerland.

C552/032/98

# Virtual reality – toy or tool?

**M HAYHOE** CEng, MIMechE
The Engineering Link Limited, Derby, UK

## SYNOPSIS

The recent privatisation of British Railways has created many new companies working within the rail industry. This demands a new culture and approach to meeting the engineering requirements of the new owners and operators.

Methods for vehicle procurement and refurbishment must now meet much more challenging timescales. Many of the franchise commitments involve such projects. These require the minimum lead time to completion in order to maximise revenue and meet agreements made.

Previously, concepts for vehicle designs have been generated through a combination of artists impressions, rough mock-up and then detailed mock-up. This has been followed by detailed design to enable procurement and eventual vehicle construction. Such a process can jeopardise the timescales required and is expensive. However, it does allow a considered decision towards what many customers regard as the most important feature of their vehicles; the interior environment.

The Engineering Link has developed a process of vehicle modelling using Virtual Reality to enable customers to develop their concepts in an active 3-dimensional world. This allows realistic options to be produced within days, at a much reduced cost compared to mock-ups. Features and colour schemes can be rapidly tried and altered to enable the customer to refine and decide on their chosen scheme. This combined with fast-track design, manufacture and construction, has supported successful refurbishment projects within the constraints demanded by the new industry.

This paper will explore the development of these techniques and the benefits that are offered for engineering projects in the railway industry.

 

# 1. WHAT IS VIRTUAL REALITY?

Virtual Reality or as it is commonly known, VR, is probably more familiar within the computer games market than commerce or industry. Those of us who have played with three dimensional (3D) interactive games, where the player can move through environments, reposition to view from a different perspective, interact with objects or simulate experience in that environment, are probably using a program developed around some VR software.

There are many definitions that can be given for VR. Stone (1) provides the following definition; "Virtual Reality refers to a suite of technologies which permit intuitive interaction with real-time, three dimensional databases".

## 1.1 Conventional Design Tools

Computer 3D representations or models are not new to the engineering industry. Many designers now work with CAD/CAE software which enables a design to be developed and represented three dimensionally. That is not to say that the traditional drawing board or two dimensional (2D) CAD design is a "dead" technique. It is often still appropriate to use these design formats, for the following reasons.

Original designs give free opportunity to consider the most effective tools to be used. Where designs are based on developments of previous products, or indeed a modification of an existing design, consideration has to be given to working with the original design medium. In the railway industry, existing vehicles can potentially last 30 to 40 years in service. At the extreme therefore, some vehicles still operating today were designed when desk top computers were the subject of science fiction. Consequently base designs are still held on tracings and conversion to electronic formats is rarely economic.

In the late 1980's, 2D CAD was used significantly in a number of multiple unit designs for British Railways. Consequently it is more common to find base designs in electronic format from this era and to develop modifications in the same format. From around 1990, components and parts of vehicle designs began to be developed using 3D CAD. Business managers could now also view the designers' work, as this format enabled rendered (i.e. showing texture, colour and form) 3D models to be displayed or printed. The concept designs for British Railway's Mark V coach were one of the first to be developed using this technique and gave impressive finished results.

However, the limitations of these 3D CAD modelling techniques need to be considered. These are summarised as follows;

- Large and expensive computing power is required to drive these models.
- 3D models of significant vehicle sections take many man-months to develop.
- Visualisation of wire frame diagrams requires an expert eye.
- Production of rendered 3D outputs is not immediate and requires much processing time.
- Interaction around the 3D output cannot be experienced.
- Design changes can be time consuming and expensive.

Demonstrations of movement about a 3D CAD model can be achieved. However, this is usually limited by the level of data and detail within the model and takes the form of a "fly through" simulation. Animation or dynamic behaviour of parts or assemblies within the model will not be possible. 3D CAD objects have to continue to store and assimilate the large amount of data that has to be input into the model. Consequently movement is usually limited by the available computing power and to models of simple components or relatively small assemblies.

For more complex models, a video can be made of snap-shots taken at changing positions, to give an impression of movement through the model. The results can be impressive in terms of observing the finished detail and rendered surfaces presented. However, this type of presentation is not interactive and therefore does not satisfy the definition of VR.

## 1.2 VR as a Design Tool

VR technology can be used on its own or to support the CAD design process. A VR model is a simple spatial representation of the world being designed. The world is built from a number of objects that are positioned to stated co-ordinates; the objects are built from facets (or polygons) that are linked together to represent shapes in the model. The data required to build this representation is far less complex than required in a conventional CAD model. Therefore, it is possible to continually deconstruct and re-build the VR world rapidly around the viewing position. Consequently real-time movement can be experienced from the model as if working in the "actual" world.

With the addition of colours and rendering of objects, the model can quickly take on an convincing representation of the completed product. Sounds, touch, movement, interaction and other attributes can be applied to the VR model. A 3D picture, image or video can be observed, but only VR can be *experienced*. It is this feature which gives VR its greatest potential as a tool for use in development and design; not just for engineers, but for business managers, marketing experts and potential operators.

The level of *experience* can be determined by the medium used to present the model. This is generally referred to as the degree of immersion. Stone (1) concludes that total immersion is yet unrealisable with present day technology. Current technology permits a range of immersion from the standard desktop PC monitor, through to the classic headset and tracker, to variants of screen projection for groups or for those with extremely generous budgets, special purpose designed VR theatres.

VR technology, as recognised today, has been around since the early 1980's. Development of VR arose from cockpit simulation work undertaken for N.A.S.A. in the USA. It is now finding increasing use in the development of manufacturing plant and production equipment. Examples of this are in the design of petrochemical plant. Here, the tortuous runs of pipework and associated equipment, located over a vast acreage of site, pose tremendous difficulties to visualise from the base 3D CAD designs tools. Any errors apparent from the design at the manufacture or installation phases of the project would not only have severe consequences to timescale, but also project cost.

VR has therefore been successfully introduced here as a design review tool. The representation and visualisation available through VR modelling enables fouls, inaccessibility and safety hazards to be quickly identified prior to physical work having commenced. The previous alternative has been to model such plant on a miniature scale (full scale is obviously impracticable!); this has been expensive, remains relatively inflexible and is not foolproof.

The automotive industry is now adopting VR in the concept and development of new cars. An objective is to reduce project time cycles. Also some companies foresee the elimination of the need for physical prototypes. The ability to experience is also being substantially developed here, with the objective of achieving full size laser produced images of the Virtual car. This is indeed at the forefront of the developing technology and is supported by substantial budgets that the automotive industry is prepared to invest against its future car production sales.

Companies such as supermarkets are using VR to model their "standard" stores. This medium enables a representation of the entire operation to be modelled from flows of stock, to positioning of checkouts, customer flows, appearance of display units and the attractiveness of the merchandise on sale. The model can act as an ideal layout and be quickly adapted to suit the particular requirements and options for a local scheme under development.

SNCF have also used VR in a research project to simulate the layout and presentation of station signage, prior to construction and implementation. The positioning and contents of signs have been modelled throughout the vast and crowded area of a station, as a means of allowing potential customers to validate the design of a future station.

## 2. VR FOR RAIL VEHICLE DESIGN

The 1980's saw the focus change within Britain's railways, to a much more business led industry. The large functional engineering departments that previously set the standards for design and maintenance were dis-aggregated into business groups. These "Business" groups and their strategies now determined the engineering performance that would be required from equipment and staff.

At the same time, this drove the consideration of business needs into the design process for any new vehicles or equipment. Industrial designers were included into the design teams and new requirements were fed in, particularly for the design of passenger and train crew environments.

### 2.1 The Design Development Process

The classic starting point for interior schemes was the development of 2D schemes from artists' sketches. This enabled the "appearance" of a product to be considered and options to be quickly and simply evaluated. The chosen options were then refined in to one or many finished impressions of various perspectives of the vehicle (e.g. saloon, vestibule, toilet, cab etc.). These perspectives could be included in a procurement specification to enable the customers' aspirations to be understood and then priced by the supplier of the vehicles.

Many of us have been brought up in a world of 2D design tools. We can look at a 2D drawing and visualise how it will look in 3D. However, the extent to which we are capable of doing this varies from individual to individual. Aspirations are therefore not always conveyed effectively by sketches and drawings. This can result in disappointment when a finished vehicle has been rolled out and does not exactly meet the expectation of the business manager who ultimately has to exploit the product.

An enhancement of this approach was therefore to build a pre-production mock-up. This enables a full size representation to be experienced as part of the design process. There are many forms of mock-up. The simple "white board" version is relatively cheap and quick to assemble. It provides an outline of shapes, positioning of features and can be altered without too much expense. It does not enable a total experience as it is lacking in texture, colour and detail. Invariably this form is a precursor to a full size detailed mock up.

The full scale mock up can come in various forms. A complete vehicle can be replicated to give a total interior and exterior representation; examples of this recently are Heathrow Express and Crossrail. More often a half or quarter size mock up is constructed to show the key features once (e.g. cab, toilet, saloon bay, vestibule) only. This reduces the size and cost of producing the mock-up, but still requires the procurement of a suitable venue for storage of the mock-up for the duration of its use.

Such models do enable the actual representation to be experienced by all and for further options to be evaluated. The degree to which options can be accommodated is however limited by the ability (and budget!) to change the details of the mock-up. The change of a seat moquette requires alternative sets of covers to be manufactured. Alternatively, the use of one or two seats only can limit the cost, but is not representative of a vehicle, when seen full of seats. The ability to quickly change panel finishes is restricted. Changes to seat positions cannot be quickly undertaken. Partition sizes and throughways cannot be changed without significant expense and time.

The human being is a very individual animal; we all have our preferences and evaluating too many options has to be restrained, when using the medium of a full size mock-up. Furthermore, this model is not easily transported or stored. This presents problems, when ensuring that individuals and organisations who require to or should be consulted during the design, have access to the mock-up at a place and time that is convenient to them and at a cost and timescale that is acceptable to the project.

The recent privatisation of British Railways has resulted in a structure of companies that are very much business led. The nature of the commitments made within franchise agreements reflects not only the delivery on financial performance targets, but also the target date milestones. The length of the franchise is sufficiently short to make critical any early decisions on investment in, or procurement of rail vehicles. Consequently, the processes previously adopted for designing rail vehicles may not fit the constraints of cost and timescales required within the new business structure.

The above process of vehicle design by sketch, preliminary mock-up, full scale mock-up then detailed design is clearly in need of review both in terms of cost and speed or ability to conclude decisions. Figure 1 illustrates the sequence, potential iterations and phases that have been discussed above. The process is nevertheless still not foolproof in evaluating the design and presents risks of on-costs or delays within any given project.

### 2.2 Development of VR Modelling for Rail Vehicle Interior Design

The Engineering Link realised that VR offered a way of improving the existing design process and satisfying the requirements from the new industry structure, by removing the cost and timescale constraints associated with physical mock-ups. In order to develop the technology, a VR model was created to represent a Mark III coach interior. This was selected because it was very much a standard vehicle structure, used on many second generation vehicles built for British Railways.

The model enabled an evaluation of the extent to which VR could be used to replace the need for a vehicle mock-up. It also became apparent to what extent vehicle options could be re-modelled and the speed at which changes could be implemented using such a tool.

Within a matter of days, a good representation of the vehicle interior was produced. Subsequent changes to seat positions, moquette covers and trim colours, could be established and evaluated within minutes. A walk through of the vehicle was also possible enabling a perspective from the passengers' view to be gained. For example, the difference from various seat head rest designs in a uni-directional layout, can inhibit a passenger's view throughout the vehicle. Preferences for seat choice could now be evaluated quickly and cheaply in this respect, using such a modelling tool.

Unlike 3D CAD, objects within the VR model could be given attributes. An example used was the power operated door from the vestibule into the passenger saloon, which was programmed to slide open or closed on the touch of the handle. A mannequin was introduced to walk down the vehicle and to use seat positions. This not only gave an impression of accessibility, but also helped in giving a perspective of scale to anyone viewing the model.

### 2.3 The Class 142 VR Model

Regional Railways North East (RRNE, now Northern Spirit) made a commitment to upgrade five Class 142 multiple units as a condition of their franchise. A business decision was taken to extend this refurbishment to thirty units, based on the projected market. The objective of the upgrade was to raise the profile of the Class 142 from a service bus to that of a motorway coach. The following areas were therefore considered;

- Wholesale replacement of the original low back seats fitted.
- Replacement heating system.
- New floor coverings and painting of interior panels.
- Replacement and improved passenger saloon lighting system.
- Conversion of one parcels area to passenger accommodation and new draught screens.
- Improved disabled persons' access
- Product advertising and livery and improved destination indicator displays

The proposed project had a number of options to be considered. The chosen options would have to meet the financial justification required against projected revenue incomes and satisfy the many interested bodies (e.g. PTE's, OPRAF) who influence and support the operations on which these vehicles are used.

Conventional prototyping was not possible with this project, since significant items (e.g. seats) were not available in production quantities. In order to maintain braking capability, designs would have to be managed to ensure that the vehicle weights remained within the existing limits. VR provided RRNE with the capability to carry out a detailed assessment of the value engineering aspects as distinct items and as a total package.

A series of models were created by The Engineering Link, using VR as the tool. A model of the existing layout and finish of the Class 142 was produced, not only as a benchmark to future changes, but to validate that VR was representative of the actual vehicle being considered. The various change options were then modelled as required by RRNE, for consideration. Figures 2 and 3 show examples of various perspectives generated during this work.

The removal of the existing draught screens from the parcels area, enabled the saloon seating layout to be re-evaluated for the entire vehicle. VR identified positions where the arrangement of seats relative to the adjacent window pillar would obscure a passenger's view. Seating positions and orientations were then optimised to avoid this problem.

Different seat types were modelled to evaluate the appearance of the saloon interior and detailed variations on seat design were also included. These could be changed or reversed wholesale throughout the saloon interior within minutes. Clearances and under-seat access with the new seats could be viewed and evaluated. This enabled access for cleaners and any potential dirt traps to be identified and considered into the design. The choices of carpet and seat moquettes were also evaluated. One particular option was for a design of carpet and seat moquette carrying the endorsement of a sponsoring business.

The design and positioning of draught screens was developed and detailed from the VR model. An illuminated logo, etched into the glass of the draught screen, could also be accurately represented and modelled. The choice and design of the main saloon lighting gondola was also included into the model. Finally and significantly, wheelchair access could be modelled and evaluated by "driving" a virtual wheelchair around the vehicle. Obstructions and fouls would readily be identified by programming the wheelchair to stop in such a situation. The new 2 plus 2 seating installation was shown to enable free access of a wheelchair throughout the saloon area of the proposed arrangement.

As stated earlier, VR is not only a model for engineers. The design can be experienced by all interested parties. RRNE used a copy of the model at their offices for evaluation within their own organisation and also to demonstrate to other bodies, such as DIPTAC, PTE's and OPRAF. This flexibility and portability was invaluable in being able to expose the proposals and evaluate them amongst the widest number of key contributors in the minimum time. The result was a commitment from the RRNE Board and their major customers, to achieving a high quality environment in vehicles which had previously been regarded as a poor product.

The model was also used to develop designs with the vehicle owners and to demonstrate the required quality, line and shape of the design to all who tendered for the final project. The ability for everyone to experience and understand the design aspirations from the VR model would ensure that there were no disappointments when the design was transformed into a production vehicle. The timescale from inception of the modelling phase to conclusion of detailed requirements was less than two months. The alternative of full scale modelling or prototyping would have significantly extended the project duration and at a substantially greater cost.

The first Class 142 unit to be refurbished on this project was completed at Heaton depot during March 1998. Work on further production units began immediately following delivery of this unit. The VR model has been exploited successfully on this project to take a concept through to detailed design and into production without the additional need for a pre-production prototype. The process has enabled the number of detailed drawings necessary to define the design safely and reliably, to be minimised. Figure 4 shows the finalised refurbished unit, which can be compared with the VR models illustrated earlier.

The finished product has been received well by RRNE, their customers, vehicle owners, and other interested parties. RRNE have commented that the finished vehicles "...look just like the model"; there have been no unpleasant surprises.

## 3. CONCLUSIONS

The technology of VR is now available to be used as a tool for the design of railway vehicles. VR gives opportunities to everyone involved in the process to ensure that the design meets business, operational and engineering requirements. VR alone will not replace and should not replace other techniques available. It does give a valuable tool that enables the design review of many options and in circumstances that would not be achievable otherwise within the timescales and costs of a project. It enables a design to be taken from VR model direct to production. It may still be appropriate to a project, for the construction of a full size mock-up, to give the ultimate experience which VR technology cannot yet provide. However, in this circumstance, VR will enable many more options to have been evaluated and optimised before the commitment to construct the full size mock up is made.

Is VR a toy or tool? It is certainly used as a toy in the games market. It definitely has been shown as a tool in the industrial market, but also is a tool that can be invaluable and fun for all to *experience*.

## 4. REFERENCES

Stone, Prof. Robert J., "A Study of the Virtual Reality Market", 1996, The Information Society Initiative for the DTI.

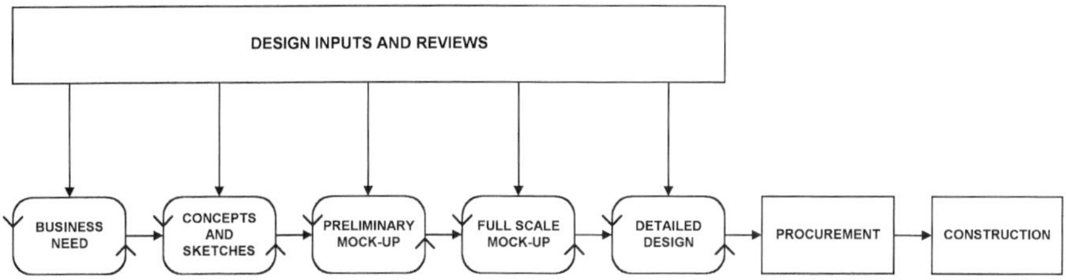


**Figure 1 The Design Cycle**

**Figure 2 VR Model of Existing Class 142 Saloon Interior**

**Figure 3 VR Model of Refurbished Class 142 Door Vestibule Entrance**

**Figure 4 Completed Refurbished Class 142 Interior**

# C552/078/98

# Progress on the Japanese Maglev system

**C J GOODMAN** MA, PhD, CEng, MIEE
School of Electronic and Electrical Engineering, University of Birmingham, UK

## 1 INTRODUCTION

A little less than a year ago, the Japanese Maglev train achieved a speed of 550 kph on the test line at Yamanashi. This represents the culmination of nearly 20 years of research and development and establishes the feasibility of the long-held ambition to provide a 1 hour journey time from Tokyo to Osaka. This paper will not attempt to review this long period of development but rather to outline the system at its current state of advancement and to present the results from the recent trials. At the heart of the Maglev system is the levitation, guidance and propulsion arrangement, all of which functions rely on the very high magnetic flux density achieved by the use of a superconducting magnet. The superconductor used is the normal low-temperature type (not the more recent high temperature superconductor which cannot sustain the current densities required) requiring cooling to liquid helium temperatures of 4° K. This, and many other aspects of such a novel system, presents some challenging operational problems which are now beginning to be tackled as the feasibility is more and more firmly established.

## 2 SYSTEM OUTLINE

The Maglev train is a passive vehicle and is propelled along the guideway in a similar manner to Eric Laithwaite's electromagnetic bullet. The guideway is a reinforced concrete structure with either separate or integral side-walls and key internal dimensions of 3.3 m×1.35 m as in Fig. 1.

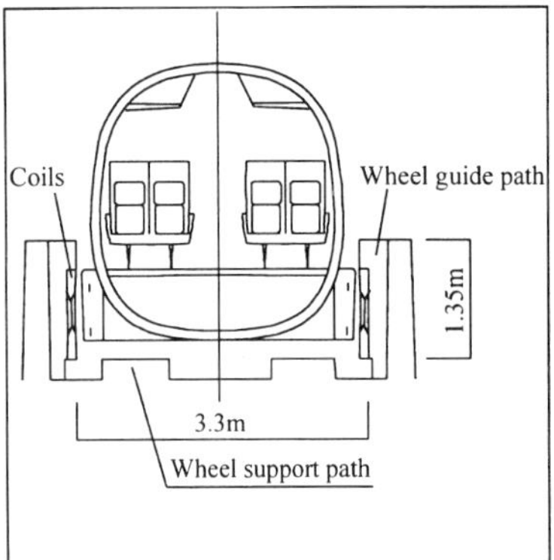

Figure 1. Basic dimensions of guideway

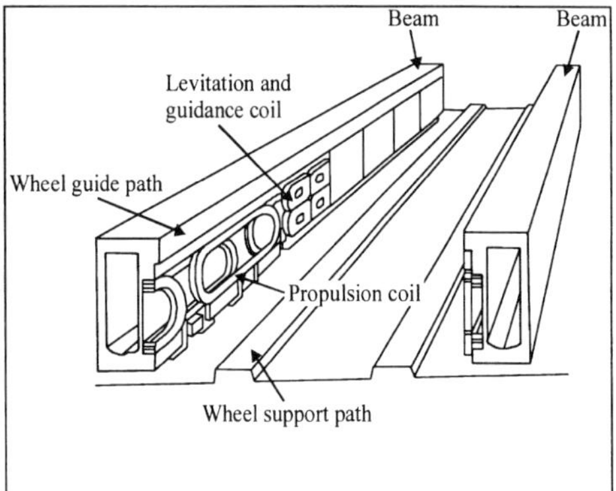

Figure 2. Side-wall beam type

At low speed (less than approximately 100 kph) the train runs on rubber-tyred wheels mounted on retractable 'landing gear', hence the surfaces provided for these. The guideway is constructed to the following main specifications:-

- Minimum radius of curves 8000 m
- Maximum slope 4%
- Maximum superelevation 10°
- Distance between guideway centre lines 5.8 m

The sidewalls also contain the propulsion and levitation/guidance coils (Fig. 2); the action of these is detailed in the next section. An inductive radio wire built into the centre of the guideway is used by antenna mounted at the front and rear of the train to measure train position to an accuracy of a few tens of millimetres. The pitch of the inductive radio wire phase change-overs is 900 mm long.

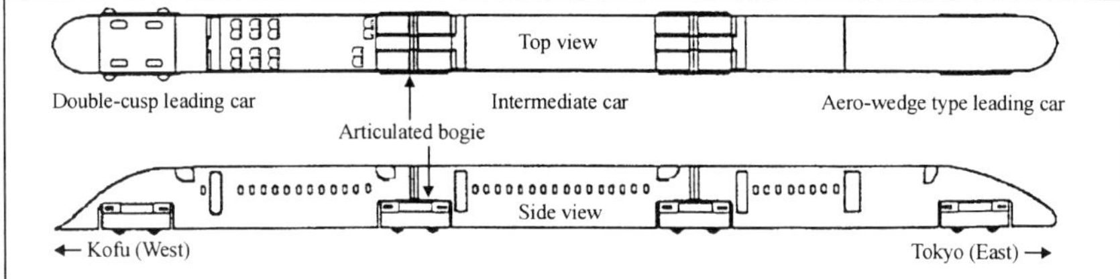

Figure 3. The first train set on the Yamanashi Maglev Test Line

The experimental test trains, known collectively as the MLX01 vehicles, consist of three cars mounted on four bogies (Fig. 3). For aerodynamic test purposes one end has a double-cusp shape and the other an aero-wedge shape. The main bogie frame supports the superconductivity magnets (SCM's Fig 4) and the primary suspension is thus actually electromagnetic with a stiffness of 5 MN/m per bogie giving a height difference of about 10 mm between empty and fully-laden. In the latest type, as fitted to train 2 and shown in Fig 4, there is an intermediate frame resiliently mounted on the SCM carrying frame; the car body is in turn supported by air springs from this intermediate frame. Earlier experiments which previous test vehicles proved this arrangement to be necessary to reduce car body vibrational accelerations as the electromagnetic suspension is very stiff and not particularly well damped.

## 3 LEVITATION AND GUIDANCE SYSTEM

The superconductivity magnet assembly on each bogie has four superconducting coils on each side. The coils are 1.07m long by 0.5 m high and built within one elaborate cryogenic system to maintain the superconductivity wire at liquid helium temperatures of 4.2° K (Fig 5). These coils are pitched at 1.35 m. Levitation is achieved passively by the currents induced into the figure-of-eight coils in the surface layer of the side-walls. The poles of the SCM induce current in such a direction that the lower coils become of the same polarity as the SCM poles and repel whereas, due to the reversal through the figure-of-eight, the upper loop becomes of opposite polarity and attracts. The nett effect is to produce an upward force so long as the horizontal centre-line of the SCM lies a little below the centre-line of the figure eights. The car height stabilises at a level just below exact alignment of the centre-lines by an amount which generates an upward force equal to the weight. Figure 6 shows the levitation characteristic in terms of the vertical displacement between the bogie and the guideway centre-lines. It is also inherent in the system that the SCM will experience a downward force if it rises above the guidance coil horizontal centre-line. Thus, a natural anti-rolling force is produced.

By cross-connecting these same coils between one side of the guideway and the other, a self-centering action is also produced electromagnetically. If the car moves closer to one side of the guideway, it experiences an increased repulsion on that side and an increased attraction on the other side, thus tending to re-centre the vehicle.

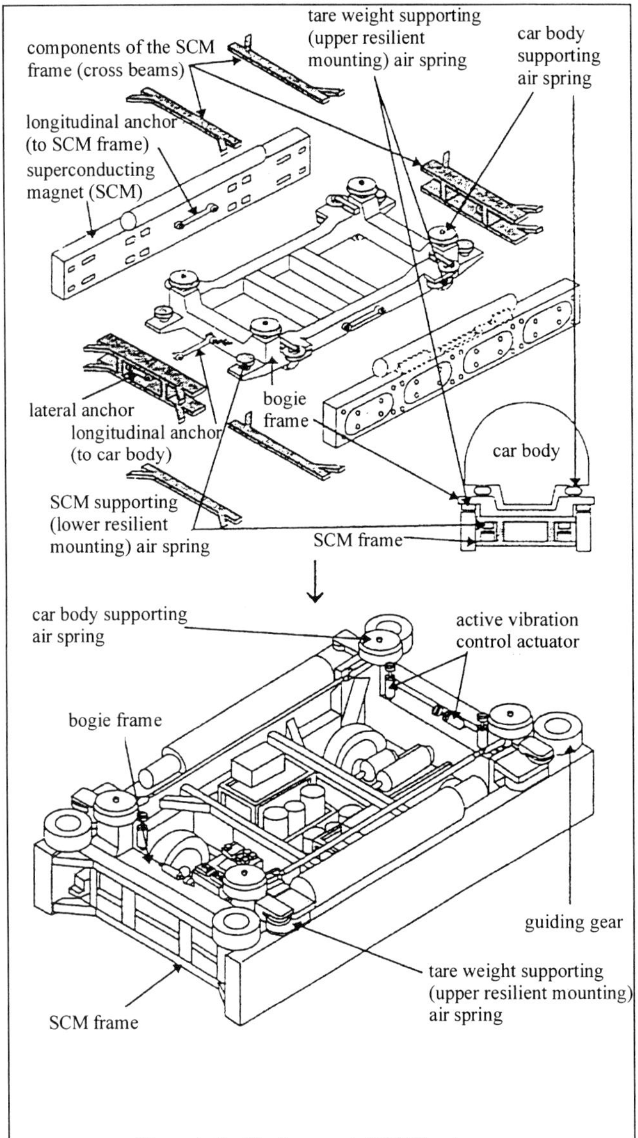


Figure 4. Resiliently-mounted SCM bogie

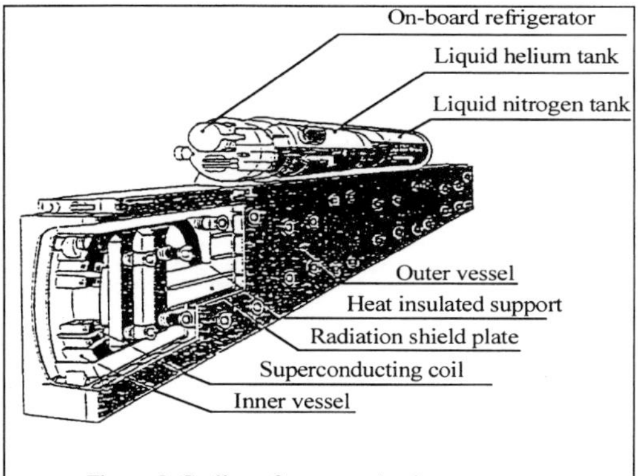


Figure 5. Outline of superconducting magnet

All of these passively generated electromagnetic forces depend on the speed of the vehicle. This is particularly true of the lateral guidance force and thus the performance of the lateral guidance is the determining factor in the speed set for lifting the landing gear. For the present tests these speeds have been set at 135 kph for straight sections and 150 kph for curved sections. The stiffness of the lateral guidance is such as to maintain a minimum 40 mm gap between the sidewall and the limiting 'stopper' wheels in normal operation. The maximum cant of 10° and minimum radius of 8000 m results in zero lateral displacement at 420 kph as shown in Fig 7.

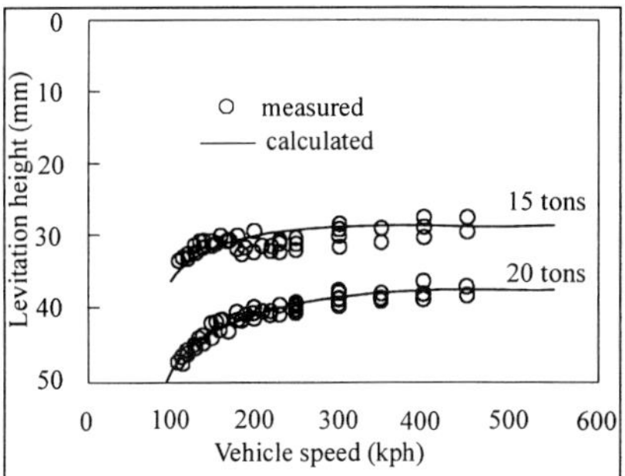


Figure 6. Levitation height vs. vehicle speed

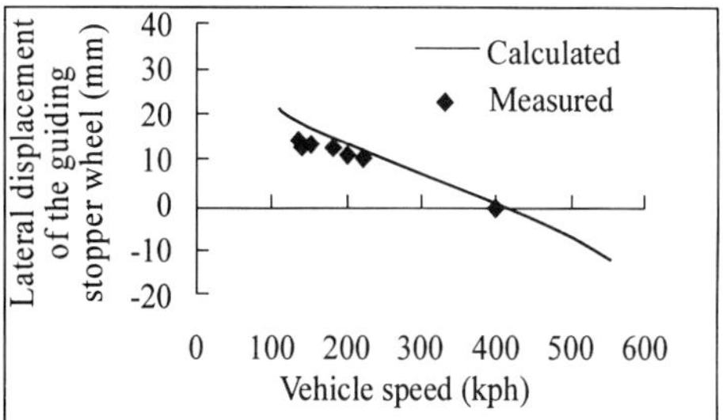


Figure 7. Lateral displacement vs. vehicle speed on a curved guideway section (cant 10°, R=8000 m)

## 4 ELECTRICAL SUPPLY AND PROPULSION EQUIPMENT

The driving force for the Maglev system is produced by the linear synchronous motor effect. The superconducting magnets on the bogies are essentially locked into a magnetic wave created by supplying variable frequency AC to the propulsion coils. These propulsion coils are mounted in the sidewalls, behind the levitation and guidance coils, in an overlapping pattern (Fig 2) and supplied with AC at frequencies between 0 and 56.6 Hz from PWM inverters at the track-side. The fundamental specifications of the linear motor are given in Table 1 below

**Table 1**

| Fundamental Specification of Linear Motor | |
|---|---|
| **Superconducting magnets (field side)** | |
| length × width | 1070 × 500 mm |
| magnetomotive force | 700 kAT |
| pole pitch | 1.35 m |
| poles/truck/side | 4 |
| **Propoulsion coils (armature side)** | |
| length × width | 1420 × 600 mm |
| disposition | 240° two-layered (every other pole) |
| pitch | 0.9 m |
| number of turns | 8 T (front side) 10 T (reverse side) |
| nominal line voltage | 22 kV |

To reduce losses, reduce the power rating necessary for the lineside equipment and provide redundancy, the power is only fed to the section of track where the train is currently positioned. The propulsion coils are grouped in sections which are staggered by half the length of a section with respect to those on the other side of the guideway. These groups are then fed in an alternate sequence using power-electronic (GTO) change-over switches from the inverters as shown in Fig 8. A key element in this arrangement is the position detection on the train. This is achieved with the 'wiggley-wire' reversed-loop type system, as already mentioned, with the loops being 900 mm long and, using phase detection, yielding a position resolution of a few tens of millimetres. The length of these coils is the same as the pitch of the propulsion coils. This same position information is, of course, used for the safety signalling.

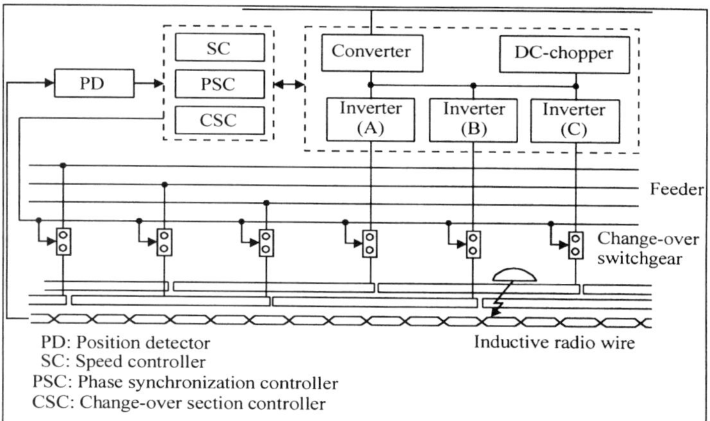


Figure 8. Configuration of northern power supply system

The power conversion chain takes in power at 154 kV which is transformed to 66 kV for the two converters, one for the Northern line and one for the Southern line as explained in the next section. The details of the converters and other equipment are given below in Table 2.

**Table 2**

| Item | Northern line | Southern line |
|---|---|---|
| Converter | | |
| Type | Thryistor converter | PWM converter |
| Capacity | 69 MW | 33 MW |
| Input voltage | 66 kV | 66 kV |
| Input frequency | 50 Hz | 50 Hz |
| Output voltage | ± 3450 V(DC) | ± 2625 V(DC) |
| PWM inverter | | |
| Number of phase | 3 | 3 |
| Capacity | 38 MVA | 20 MVA |
| Output current | 0-960 A | 0-1015 A |
| Output voltage | 0-12700 V | 0-6350 V |
| Output frequency | 0-56.6 Hz | 0-45.3 Hz |
| Carrier frequency | 500 Hz | 300 Hz |
| Element (GTO) | 4500 V-3000 A | 45000 V-4000 A |
| Element comp. | 4S-1P-4A/unit | 3S-1P-4A/unit |
| cooling type | water cooling | water cooling |
| DC chopper | | |
| Capacity | 19 MW | 4.09 MW |
| Input voltage | ± 3550 V | ±2625 V |
| Chopping freq | 300 Hz | 350 Hz |
| Circuit comp | 5 parallels | 3 parallel |
| Control method | PWM control | PWM control |

The converter on the northern side is a 24-pulse thryistor converter and equipped with Static VAr Compensators and harmonic filters. From the DC link three three-phase PWM inverters are operated and a DC chopper is also provided to absorb regenerated energy since the linear synchronous motor inherently provides the main braking in normal operation. Emergency braking is achieved by short-circuiting the propulsion coils and using air-brakes. At lower speeds, disc-brakes are available on the landing wheels. On the southern side, the converter is a PWM GTO converter allowing the regenerated energy to be passed back into the main AC supply. Nevertheless, a DC chopper is also provided, of less capacity than the northern side.

The inverters consist of four units transformer-coupled in series to obtain the high voltage with low harmonies. As this arrangement is unsuitable for producing low voltages at low frequencies one of the inverters can be used separately in these conditions. The changeover is made whilst the inverter is 'off-line' between propulsion sections.

The drive is controlled with a closed-loop speed controller which takes its reference from a pre-calculated speed reference pattern based on desirable performance and ride features such as jerk rate, acceleration and deceleration values. The 'error' between the measured speed, calculated within the position measurement system, adjusts the phase and frequency of the inverters so that the vehicle thrust is adjusted to reduce the error.

## 5 THE YAMANASHI TEST LINE

The Yamanashi test track is located about 100 km west of Tokyo and consists of a principal section 18.4 km long (the northern line) paralleled for 12.8 km by the southern line thus enabling passing tests to be conducted. A location map is shown in Fig 9. Of the 18.4 km, 16 km are in tunnel and 2.4 km are open. A test centre has been built on an open section about one-third of the way along the line from the eastern end; one of the power feeding sub-stations is also located here. This facility has been operational since 1996 and is the first part of a planned 42 km line.

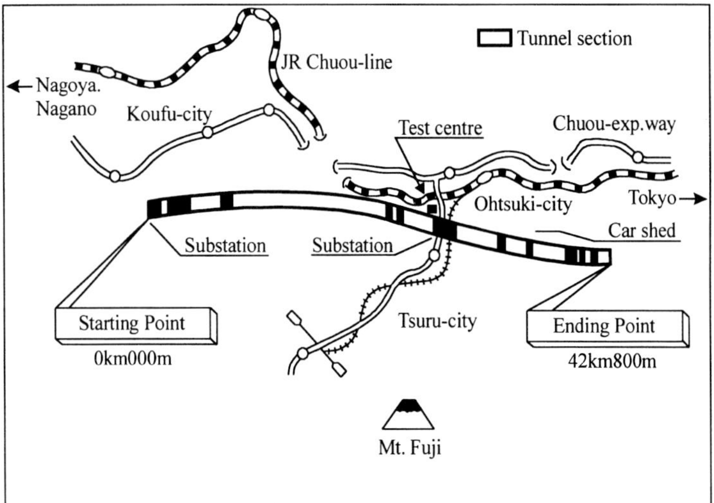


Figure 9. Location of Yamanashi test line

Two train sets are available; the first was delivered in July 1996 and the second in October 1997. The main specifications are given below in Table 3.-

## Table 3

| Category | | First train set | | | Second train set | | | |
|---|---|---|---|---|---|---|---|---|
| | | Leading car (Xofu) MLZ01-1 | Middle car MLX01-11 (Standard) | Leading car (Tokyo) MLX01-2 | Leading car (Xofu) MLX01-3 | Middle car MLX01-21 (Long) | Middle car MLX01-12 (Standard) | Leading car (Tokyo) MLX01-4 |
| Maximum Speed | | 550 kph | | | | | | |
| Pole pitch | | 1.35 m | | | | | | |
| Composition | | Superconducting Magnet (SCM) concentration, Articulated truck | | | | | | |
| Structure | | Aluminium alloy semi-monocoque structure | | | | | | |
| | Maximum width | Car body: 2.9 m Truck: 3.22 m | | | | | | |
| | Cross section | 8.9 $m^2$ | | | | | | |
| Dimensions | Height | Levitation: 3.28 m (on-gear running: 3.32 m) | | | | | | |
| | Vehicle Length | 28.0 m | 21.6 m | 28.0 m | 28.0 m | 24.3 m | 21.6 m | 28.0 m |
| | Train Length | 77.6 m | | | 101.9 m | | | |
| Height (Loaded) | | Leading car: 32 t Middle car (standard): 20 t | | | Leading car: 33 t Middle car (Standard: 20 t Long: 22t) | | | |
| Nose shape | | Double cusp | - | Aero wedge | Aero wedge | - | - | Double cusp |
| Passenger compartment | Capacity<br>Seats per row<br>Set pitch | 46 seats (2+2 per row)<br>0.88 m | - | 30 seats (2+2 per row)<br>0.88 m | - | 70 seats (2 + 2 per row)<br>0.88 m | - | - |
| Others | SCM | 4 poles per side. Same polarity on right and left sides (700 kA) | | | | | | |
| | On-board brake | Aerodynamic brake + Wheel disc brake | | | | | | |
| | Body colour | Base: oyster white, Line: Navy blue | | | | | | |
| | Fire prevention | Application of Railway Structure Regulations, Non-combustible and flame-resisting materials, Fire alarm | | | | | | |
| | Main equipment | . Crew cabin<br>. On-board controller | - | . On-board power supply | . Crew cabin<br>. On-board controller | - | - | . On-board power supply<br>. Inductive power collection |

An outline of the test schedule is given below in Table 4. This naturally includes some history and some prediction. The results from tests already conducted will be reviewed in the next section.

**Table 4**

| Year & Month / Test Items | 1997 (April – March) | 1998 (April – March) | 1999 (April – March) |
|---|---|---|---|
| Basic running test | . General adjustment test<br>. Speed acceleration test<br>. Maximum speed verification test | | |
| General functional tests | | . High speed differential test<br>. Multiple-train control test<br>. Various functional test | |
| Reliability verification tests | | . Transportation capability verification<br>. High speed continual running test | |
| Other tests | | . Passenger physiology confirmation test<br>. Environmental impact and environmental impact countermeasure verification test | |

## 6 TEST PERFORMANCE

The most high-profile results to appear from the tests at Yamanashi are naturally the speeds achieved. These speed trials are only one aspect of an immensely detailed sequence of tests and could only be attempted after a programme of wheel-running tests and levitation characteristic tests; some results from the latter have already been given in an earlier section. The speed trials were conducted during the latter part of 1997 with the speed gradually being raised in 20-30 kph steps. The milestones of this series of tests were as follows:-

- 421 kph 12/9/97 (manned)
- 451 kph 3/10/97 (manned)
- 503 kph 28/11/97 (manned)
- 531 kph 12/12/97 (manned)
- 550 kph 24/12/97 (unmanned)

This last result is a world record and represents the design maximum speed of the vehicles, with 500 kph being the top operational speed in service. By the end of December 1997, the culmative distance logged in running tests exceeded 11000 km.

Although the speeds achieved are of major significance, a large number of other items were simultaneously under test, the main ones being:-

- Disc brakes/aerodynamic brake/rheostatic and short-circuit coil brake
- Lineside environmental impact
- High speed landing
- SCM vibration characteristics
- Pressure distribution on car body
- Running resistance
- Air pressure damping effect of tunnel hood
- Air disturbance in adjacent guideway
- Magnetic field on passenger platforms
- Millimetric radiowave transmission characteristics
- Sidewall dynamic response

The following sub-sections give brief details of some of the test experiences

## 6.1 Drive characteristics

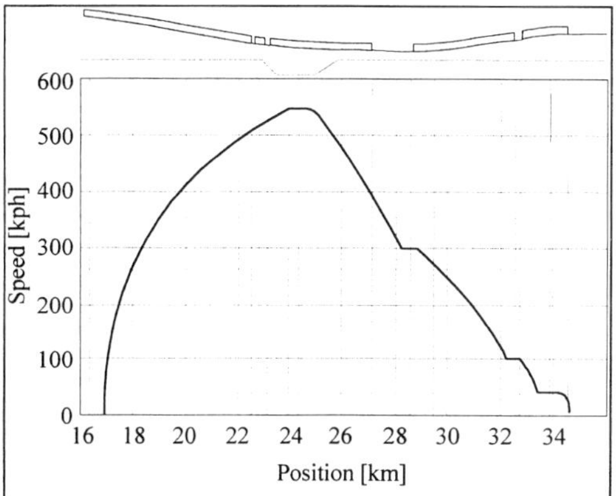


Figure 10. Run curve reaching 550 kph

The drive control system was confirmed to operate stably and, up to 350 kph, an acceleration of 0.2 g is achieved. The regenerative brake is also stably controlled with a deceleration of 0.25 g being achieved. A curve from the record speed trial on 24/12/97 is given in Fig 10 which shows the well-controlled performance.

### 6.2 Brake performance

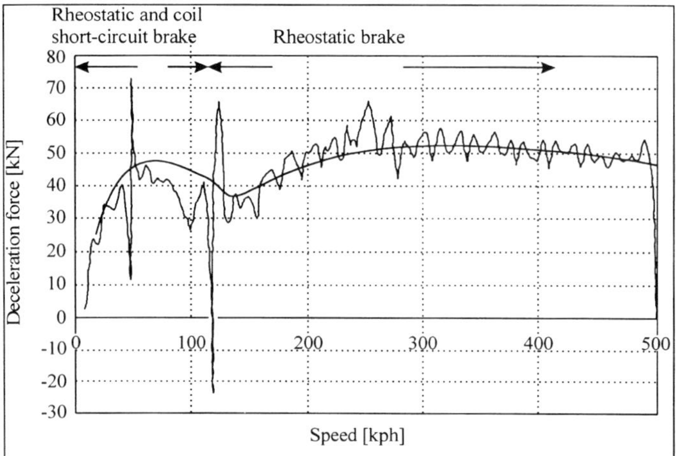

Figure 11. Electric safety brake characteristics

The main brake is intrinsically regenerative electric. Back-up braking is provided first by rheostatic braking, then by shorting the coils at speeds below 140 kph. If these primary safety braking systems fail, the two mechanical systems, air-brake and disc brake can be used with the latter coming in at speeds below 300 kph. Fig 11 shows a test recording of the braking force achieved by the primary electrical safety brake.

The aerodynamic brakes have also been tested, at speeds up to 460 kph so far, and the performance has exceeded expectations. Tests have been conducted both in open and tunnel sections. Disc brake tests have shown that with anti-skid control, the disc brakes can successfully produce 0.15 g deceleration at speeds below 300 kph.

### 6.3 Vehicle dynamics

Run-down tests in both open and tunnel sections have been conducted with results showing the running resistance to be a little lower than predicted by calculation and model tests. The results are slightly different in the two tunnels presumably because the detailed guideway structure is different. The levitation and guidance characteristics, as previously discussed, have also been carefully measured and found satisfactory. The SCM assemblies are potentially subject to damage by vibration and thus a series of detailed vibration measurements have been conducted to prove that these key components can operate stably in service conditions.

### 6.4 Environmental Impact

In early tests, as the speed exceeded 400 kph some problems were experienced with air pressure disturbances affecting nearby houses, especially those near the tunnel entrances. This was traced to two effects; entering the tunnel and passing holes for equipment in the tunnels. The latter problem has been substantially cured by plugging 13 out of the 27 holes.

The tunnel-entering problem is actually at the opposite end where an explosive sound can be generated. Air pressure damping hoods at the tunnel entrances can reduce this effect substantially.

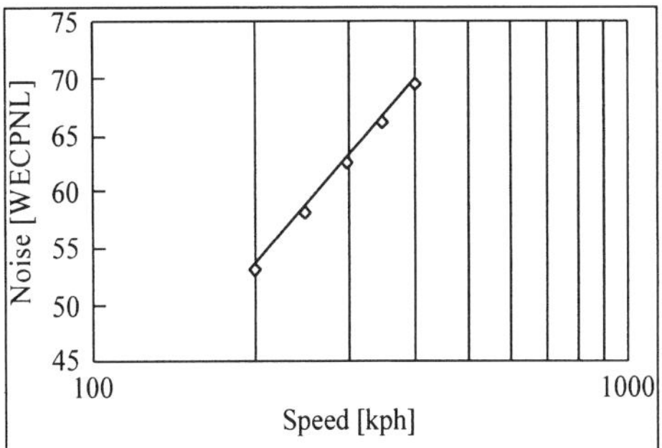

Figure 12. Running noise level

A running noise level of 75 dB according to a standard measuring 'weighted equivalent continuous perceived noise level' is the maximum permitted. This has to be measured at 25 m from the guideway and 1.2 m above ground level. Fig 12 shows the test results at up to 400 kph from which it can be predicted that the target will be met at 500 kph.

## 7 THE FUTURE

With two train sets now in operation the crucial testing of high-speed passing and of one train passing a stationary train at a platform can be undertaken. Other functional tests will involve a train crossing over between substation areas, and multiple train operation. A system for inductive power collection for hotel power on the train will also be tested together with long-term reliability tests.

The second train has the resiliently mounted SCM and the characteristics of these need to be proven.

Meanwhile, the ridership on the Tokaido Shinkansen linking Tokyo to Osaka on a route essentially along the coastal plains has reached 367,700 passengers per day and it is expected that his line will not be able to provide sufficient transportation capacity for the demands of the 21st century. Thus even though the 'Nozomi' train-sets have achieved a significant reduction in journey time and Shinagawa Terminal will further increase capacity, the Tokaido Shinkanen will reach a limit of speed and capacity. Hence the Japanese Ministry of Transport has identified a route through the mountains referred to as the Chuo Shinkansen.

By the end of the tests on the Maglev test line at Yamanashi in the year 2000 it is expected that the feasibility, reliability and service suitability of the Maglev System will thus be a major contender for operation on the new Tokyo-Osaka alignment.

## 8 ACKNOWLEDGEMENTS AND BIBLIOGRPAHY

It is in the nature of this review paper that information has been drawn from many sources and that referencing in the normal way is not particularly useful. The author would, however, like to acknowledge unreservedly the great contributions made by all the staff of the Central Japan Railway Company and the Railway Technical Research Institute in Japan who have been involved in the Maglev project. He is particularly endebted to Dr A Seki, Central Japan Railway Company, who has checked the paper in great detail.

For specific elaboration of many of the points made in this paper, the following papers from the Maglev '98 conference held at Yamanashi in March 1998 will be found invaluable.

i) 'Development of the Maglev system in Japan: Past, Present and Future; Hiroumi Soejima, RTRI and Katsutoshi Isoura, CJR, pgs 8-11.

ii) 'Superconducting Maglev Development in Japan'; Hiroshi Nakashima, RTRI and Katsutoshi Isoura, CJR, pgs 25-28.

iii) 'Outline of Yamanashi Maglev Test Line and Test Schedule'; Yutaka Osada, CJR, Kazuo Sawada and Fuminao Okumura, RTR, pgs 50-55.

iv) 'Current test status of the superconducting Maglev System on the Yamanashi Test Line'; Hisana Kurobe and Kan-ichiro Kaminishi, RTRI, Shigeki Miyamoto and Akio Seki, CJR, pgs 56-61.

v) 'Guideways and structures on the Yamanashi Maglev Test Line and their dynamic response characteristics'; Mikio Yamazaki, Yasuyuki Gotou and Shigeo Aoki, CJR, Shoichi Hashimoto and Masamichi Sogabe, RTRI, pgs 178-183.

vi) 'Train Control characteristic and the Function of the Position Detecting System at Yamanashi Maglev Test Line'; Motoharu Ono, Yasushi Sakuma, Hisanobu Adachi and Kenji Hanamotto, CJR, Toshikatsu Watanabe, Yoshifumi Sakai and Shin Sasaki, RTRI, Yoshinao Yokota, Nippon Signal, Sadao Uchida, Kyosan Electric Mfg Co., pgs 184-189.

vii) 'Results of Running Tests and Characteristics of the Dynamics of the MLX01 Yamanashi Maglev Test Line Vehicles'; Hiroshi Yoshioka, Erimitsu Suzuki, Hiroshi Seino and Masayoshi Azakami, RTRI, Hiroshi Oshima and Toshikatsu Nakanishi, CJR, pgs 225-230

viii) 'Characteristics of Superconducting Magnets and Cryogenic System on Yamanashi Test Line'; Hideo Tsuchishima, Takashi Mizutani, Tadao Okai and Masahiko Nakauchi, RTRI, Motoaki Terai, Satoru Inadama and Tetsuo Asahara, CJR, pgs 237-243

ix) 'Driving Control Characteristic Using the Inverter System at Yamanashi Maglev Test Line"; Makoto Hashimoto, Jun-ichi Kitano, Kaoru Inden and Hioaki Tanitsu, CJR, Ikuo Kawaguchi, Shigeo Kaga and Tsuyoshi Nakashima , RTRI, Shigeki Koike, Hitachi, Yasuyoshi Mugiya Toshiba and Hideki Sogihara , Mitsubishi, pgs 287-291

# Authors' Index


## B

Barrow, H ........................................ 53–66

## C

Catling, D T .................................... 43–52
Chaplin, A J .................................... 33–42

## D

De Geest, E ...................................... 21–30

## G

Goodall, R M .................................... 67–76
Goodman, C J ............................. 115–128
Griffin, T ......................................... 43–52

## H

Hardy, A E J ....................................... 1–12
Hayhoe, M .................................. 105–114

## L

Lilley, D M ....................................... 77–86

## M

Manning, C J .................................... 13–20
McCulloch, C F ................................ 21–30
Mylroi, I R ....................................... 33–42

## P

Perrott, F C ...................................... 87–94
Pope, C W ....................................... 53–66
Pope, D J ......................................... 53–66
Price, M J ........................................ 33–42

## R

Renfrew, A C ................................... 87–94

## S

Scott, J W ...................................... 95–104

## W

Walker, A J ..................................... 33–42

The Institution of Mechanical Engineers is a leading forum for the exchange of knowledge and expertise in the field of mechanical engineering.

A wide range of events is organized by the IMechE, to which all are welcome to attend. For further information about IMechE Conferences, Seminars, Workshops, and other events please contact us for further information.


| | |
|---|---|
| Visit our website | www.imeche.org.uk |
| Telephone | +44 (0) 171 222 7899 |
| Fax | +44 (0) 171 222 4557 |
| Or write to | Conferences and Events<br>Institution of Mechanical Engineers<br>1, Birdcage Walk<br>London<br>SW1H 9JJ |


We look forward to seeing you at our future events.